# Computational Music Science

More information about this series at http://www.springer.com/series/8349

Guerino Mazzola • Alex Lubet • Yan Pang
Jordon Goebel • Christopher Rochester
Sangeeta Dey

# Making Musical Time

Springer

Guerino Mazzola
School of Music
University of Minnesota
Minneapolis, MN, USA

Alex Lubet
School of Music
University of Minnesota
Minneapolis, MN, USA

Yan Pang
School of Music
University of Minnesota
Minneapolis, MN, USA

Jordon Goebel
School of Music
University of Minnesota
Minneapolis, MN, USA

Christopher Rochester
School of Music
University of Minnesota
Minneapolis, MN, USA

Sangeeta Dey
School of Music
University of Minnesota
Minneapolis, MN, USA

ISSN 1868-0305 ISSN 1868-0313 (electronic)
Computational Music Science
ISBN 978-3-030-85628-1 ISBN 978-3-030-85629-8 (eBook)
https://doi.org/10.1007/978-3-030-85629-8

Mathematics Subject Classification (2020): 18Bxx, 18Fxx, 00A65, 97M80



This Springer imprint is published by the registered company Springer Nature Switzerland AG
The registered company address is: Gewerbestrasse 11, 6330 Cham, Switzerland

*Dedicated to Emilio Lluis-Puebla*

Photo and © by Luis Solórzano

# Preface

It is time now to talk about time, more precicely: time in music. Why? Because within the complex topic of time, musical time plays a very special role. It is not identified as a type of time, which we encounter in the general existentiality of humans, but reveals a constructive character that transcends the general context and proves also a precise mechanism which represents more than a passive perception of a given ontology as often and typically conceived in physics and the neuroscientific cognitive perspective.

Such a project must first of all deal with the concepts of time which are set forth in overarching contexts of time theories. We are forced to elaborate the musical time reality as a specific expressivity that differs from non-musical ontologies. Such a project has to focus on musical time as a very special form that relates to other conceptualizations, and to elaborate its distinctive and characteristic attributes. But our analysis of time in philosophy, physics, and cognitive sciences reveals a number of important commonalities and roots, which put into evidence a convergence of time concepts, which induce its musical ontology.

This book follows a set of approaches to music theory, and more precisely: mathematical music theory, which include our universal setup as developed in *The Topos of Music* [94, 95, 96, 97], the discourse about gestures in *Flow, Gesture, and Spaces in Free Jazz* [99], the analysis of *Musical Creativity* [101], and the theory of *Musical Performance* [100]. These theories all deal with time, but never set the focus onto analyzing and understanding the in-depth ontology of musical time.

In this book, we now elaborate a temporal ontology of musical time that follows an explicit theory of how musical time is constructed. We contend that *musical time is a construction*, this is our first basic thesis. But we also propose and develop the idea that *musical time is constructed from a sophisticated interaction of musical gestures.* Musical time is presented as a result of a "distributed identity" that appears from a mathematically conceived mechanism involving projective limits of gestural diagrams.

This endeavor is developed not only from the mentioned philosophical, physical, and cognitive perspectives, but also from existing approaches to musical time as developed by well known musicological models. In particular, we analyze Jonathan Kramer's creative concept of a vertical time, which will be embedded in our model that comprises imaginary time, which was proposed by Stephen Hawking and Izak Bars in cosmology.

We then apply our model to several experimental compositional approaches that are being realized by composers Yan Pang, Jordon Goebel, and Chris Rochester. We also acknowledge Renan Madeira's precious comments concerning Merleau-Ponty's perspectives and South-American time cultures.

The cultural spectrum of musical time proves the dominance of a gestural genealogy through many important geographic and historical loci. The delicate construction of musical time from its gestural mechanisms is discussed and made evident in a subtle professional analysis of handedness by Alex Lubet in the performance of string instruments. The deep tradition of Asian time cultures is covered beautifully by Yan Pang (Chinese/Japanese approaches) and Sangeeta Dey (Indian approaches).

This book is an interdisciplinary perspective, but a necessary one that makes evident the central role of musical time construction in human existence. Musical time is probably the most human achievement, and we claim that our model sets forth a precise and realistic model of this substantial creation.

We have abstracts for chapters and sections, represented in short "summary" paragraphs. We also add concluding paragraphs at the end of chapters and sections, under the title of "impact", where we connect the contents to other sections or chapters, and, most importantly, draw a vector to the global thesis of this book that musical time is a gesturally embodied construction.

We are pleased to acknowledge the strong support from Springer's science editor Thomas Hempfling for writing such a demanding treatise in covid19 pandemic.

Minneapolis, October 2020

Top left to right, zoom view: Guerino Mazzola, Yan Pang, Jordon Goebel.
Bottom left to right, zoom view: Christopher Rochester, Alex Lubet, Sangeeta Dey.

# Contents

## Part V Conclusions

# Part I

# Ontological Orientation

# 1

# Ontology, Oniontology, and the Artistic Presence

**Summary.** This short chapter introduces first the global architecture of ontology of music, which this book is going to use extensively. It then presents an ontological model of artisic presence, which is an essential background architecture for the construction of musical time.

– $\Sigma$ –

## 1.1 Ontology and Oniontology

This section is about ontology of music, including three dimensions: realities, semiotics, and communication. It also includes the extension of ontology to the fourth dimension of embodiment. We call this extension "oniontology" for reasons that will become evident soon.

## 1.2 Ontology: Where, Why, and How

Ontology is the science of being. We are therefore discussing the ways of being that are shared by music. As shown in Figure 1.1, we view musical being as spanned by three 'dimensions', i.e., fundamental ways of being. The first one is the dimension of realities. Music has a threefold articulated reality: physics, psychology, and mentality. Mentality means that music has a symbolic reality, which it shares with mathematics. This answers the question of "where" music exists.

The second dimension, semiotics, specifies that musical being is also one of meaningful expression. Music is also an expressive entity. This answers the question of "why" music is so important: it creates meaningful expressions, the signs that point to contents.

The third dimension, communication, stresses the fact that music exists also as a shared being between a sender (usually the composer or musician), the



G. Mazzola et al., *Making Musical Time*, Computational Music Science,
https://doi.org/10.1007/978-3-030-85629-8_1

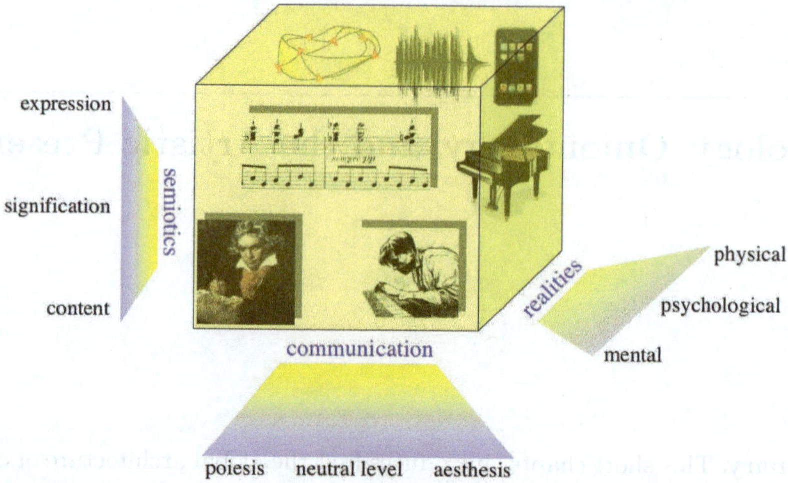


Fig. 1.1: The three-dimensional cube of musical ontology. ©Guerino Mazzola

message (typically the composition), and the receiver (the audience). Musical communication answers the question of "how" music exists.

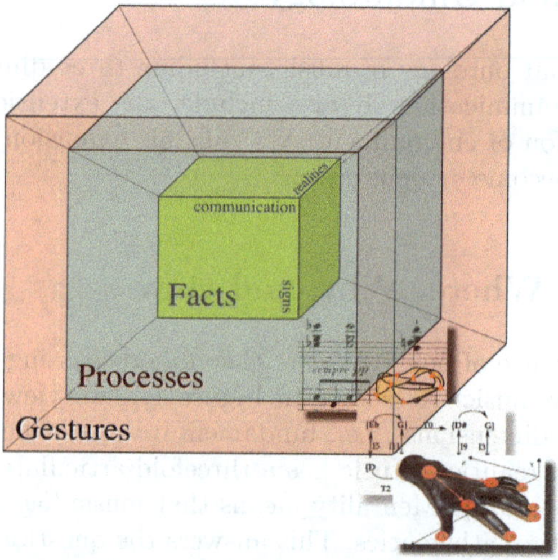


Fig. 1.2: The hypercube of musical oniontology. ©Guerino Mazzola

## 1.3 Oniontology: Facts, Processes, and Gestures

Beyond the three dimensions of ontology, we have to be aware that music is not only a being that is built from facts and finished results. Music is strongly also processual, creative, and living in the very making of sounds. Musical performance is a typical essence of music that lives, especially in the realm of improvisation, while being created. The fourth dimension, *embodiment*, deals with this aspect; it answers the question "how *to come into* being?" It is articulated in three values: facts, processes, and gestures. This fourth dimension of embodiment gives the cube of the three ontological dimensions a threefold aspect: ontology of facts, of processes, and of gestures. This four-dimensional display can be visualized as a threefold imbrication of the ontological cube, and this, as shown in Figure 1.2, turns out to be a threefold layering, similar to an onion. This is the reason why we coined this structure "oniontology"—it sounds funny, but it is an adequate terminology.

## 1.4 A Short Characterization

The four dimensions of musical oniontology can be put together to present a short characerization of music:

Music embodies
meaningful communication
and
mediates physically
between its
emotional and symbolic layers.

## 1.5 The Artistic Presence: A Processual Unfolding of Oniontology

Although we shall develop the technical details of structure and rationales of time performance later in this book, it seems important to draw a compact image of the existential focus, which is embraced by an engaged performer. There are two reasons for this preliminary setup:

1. We would like to sketch an aspect of time in performance that is vital to the understanding of the living artist, and
2. we want to give a first hint to every performer who isn't happy with a merely technical understanding of his/her artistic life. It is only a germinal hint, since the existential shape of a performing artist is far from being understood, and it is not the target of this book to complete research on this complex topic. We are, however, targeting the temporal aspect thereof.

Let us briefly digress on the very difficulty of such an enterprise and have a look at the philosophy of dance as it has been addressed by Paul Valéry in [160].

A dancer is a performer *par excellence*, since the dancer's score, written in Laban notation or any other dance score language, is known to be a poor reference to dance in action. This was analyzed in detail in [55]. Valéry in his treatise recognizes that dance is more action than every other art, that it is an art of time, and that it is an art in which the artist's life is taken in its full extension as the dancer's body is fully engaged in the unfolding of dance.

But then, instead of proceeding to a valid definition of dance (what he explicitly wants to achieve), he looks into the philosopher's mirror and recalls that Aristotle, Nietzsche, and in fact all philosophers are dancing with their words and thoughts. What a pirouette of thought: Valéry, instead of writing a philosophy of dance, makes philosophy dance. This dancing philosophy seems to miss the point: You usually do not represent knowledge about an object by making it. Valéry's approach does not give the expected definition of the concept of dance. However, he says that dance, as an art of time, pertains to a fundamental quality of human existence, namely time. And he recalls that Saint Augustin admitted that he knows what time is, but when asked about time cannot define it. Dance for Valéry is a similar phenomenon: Impossible to define, you have to live it, and he lets us know that thinking is a way of dancing.

This insight might seem precious, but we cannot accept the answer "Let's dance, and you will know!" to the question "What is dance?" Of course, the answer reminds us of that famous saying that dancing is a way of thinking. And for our concerns it is also true that performing music is a way of thinking. But here lies the problem: If it is a way of thinking, in what sense is it a special way, and different from other ways of thinking? The performer thinks music in very specific ways, radically different from the way a music theorist thinks music—see Martin Puttke's paper "Learning to dance means learning to think!" in [130]. So let us try to describe those characteristic coordinates of this way of thinking music.

It is a logical necessity to locate the performer's (oni)ontology in the framework of multi-agent communication since a performer should focus on communication among *all* agents. A performing musician cannot limit his/her interaction to a unidirectional messaging from composer to audience. This is an outdated casting of performative creativity as slavish service for the ingenious composer, an all-too-narrow perspective propagated by Arnold Schönberg and resonating in a casted performance education that produces only robots, not musicians. It is a sad fact that even in standardized jazz education, the slavish messaging of jazz on the basis of lead sheet changes replaces creative interaction of improvisers and disseminates that "whitened jazz" catechism, against which—apart from more general social and political motives—free jazz in the 1960s of the 20th century was rebelling.

Within the three dimensions of musical oniontology that complement the communicative dimension, namely realities, embodiment, and semiotics, the performer realizes a crossing of a singular type. In what follows, we stress the characteristic features (which does not mean that other aspects are absent).

The most characteristic feature on the axis of realities is the interaction of two bodies: the musician's body and the body of physical sounds. Their interaction is generated on the interface of the musical instrument, whose bodily manipulation produces the music's sounds. For an acting performer, this coupling of bodies is the core neutral niveau. All other levels of neutrality might be implied or subsumed, but this one is the manifest neutral building block. The confrontation of the musician's body with the body of sound along the instrumental interface is a factual reality: this is *the resulting fact* of a performer's action, it answers the question of *WHAT?* is being the fact.

On the axis of embodiment, corresponding to the reality of instrumentally interacting bodies, the performer's focus is on gestures. It is these gestures that are communicating musical formulas or processes. It is the highest quality of musical expressivity to deploy compact musical formulas into gestures. It answers the question of *HOW?* the facts are being produced. Gestural embodiment does not populate given spaces but creates them, defines their extension and thereby enables fellow musicians' gestures to resonate with one's gestures in these shared spaces. "Understanding is catching the gesture and being able to continue." This deep insight by the French philosopher and mathematician Jean Cavaillès [18] is what happens in the gestural interaction among performers: Their gesturally spaced vibration is what drives their bodies to move and to shape the "body of time."

Relating to semiotics, the (successful) performer organizes the future of the music being performed with reference to the music's past. The meaning of the music played to this present moment is connected to the shaping of the meaning of the next musical signification in a flow of thoughts. Such a transfer of structures is a processual item, it connects the gestural vibration and the factual display of bodies. It answers the question of *WHY?* the gestures are being produced to generate the factual output.

In order to achieve this in a coherent and persuasive way, we have to identify an environment where such a strong shaping activity is executed. On the level of physical events, we cannot realize such a program, since physical time presence is a real number $t_0$, past is the interval $t < t_0$, and future is $t > t_0$. Therefore classically conceived physical presence reduces to a single time point, and nothing can really happen in such a vanishing point.

This perspective is not satisfactory from the performer's point of view since the concept of presence in the time-sensitive arts cannot be reduced to zero. We do not embark in a neurophysiological model of performer's presence, because to our knowledge there is no such model. We rather want to postulate such a reality of artistic presence independently of a neurophysiological modeling. The fact is that the time-space of presence in artistic creation and shaping of structures is a huge environment where all the logical decisions upon performative actions are made: gestural strategies, receiving and processing the structures of past musical events, the contributions from other agents in the multi-agent network, the knowledge from the symbolic score and its prefabricated analyses, etc. We need such a time-space that conceptually is independent of the physical time point of presence. Let us call this time-space "imaginary" for two reasons. On the one hand, it is an environment that pertains to the psychological reality and as such is imagined. On the other, and this is a speculative thought, it is known that modern physics—in the research of Stephen Hawking [54] in Big Bang cosmology, and also in contributions of Itzhak Bars [12] to the unification of gravitation and quantum mechanics—has introduced a second time dimension. We shall discuss complex time in depth in Chapter 12.1.

Offering to the artist the concept of imaginary time-space is a completely different affair: It opens a huge environment where the artist's consciousness can evolve and construe complex architectures and highways of shaping musical structures. The gestural complex is driven in this time-space; the flow from past to future music that is being played is driven and lived in this imaginary realm. The overall image is drawn in figure 1.3.

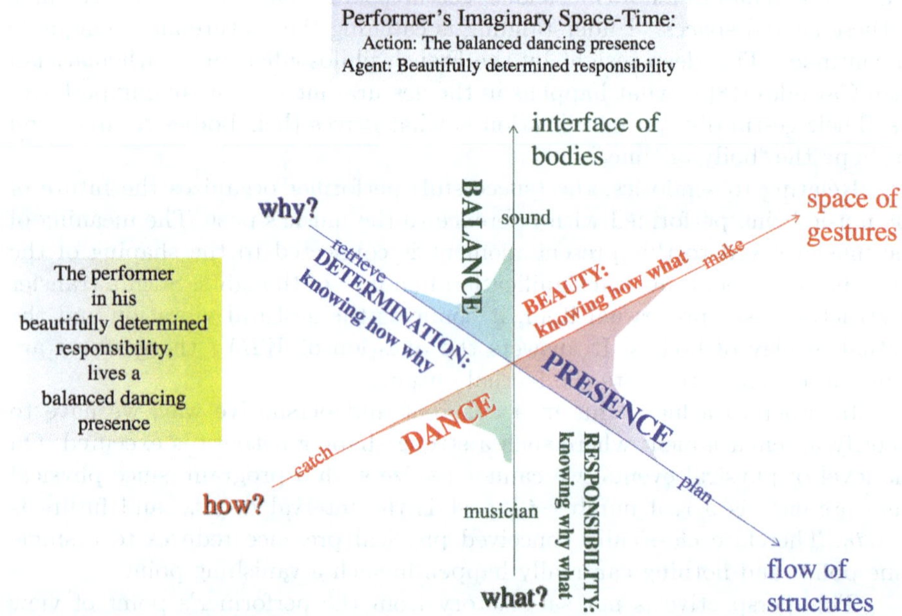


Fig. 1.3: One could conceive musical performance as being the balanced dancing of presence, an existential intensity in three directions, facts (bodies), processes (structural transport), and gestures (gestural interaction), of communicative activity. ©Guerino Mazzola

Let us discuss this image in detail. Let us first focus on the processual flow process connecting past music instances to the shaping of future ones. This aspect can take place in very different ways. To begin with, the performers may rely on different extensions of the past music's structure. An improviser might only listen to one other fellow musician or he/she might listen to many of them when shaping the next sounds. But these future sounds might also be shaped by a more or less strong reference to pre-conceived structures, independently of the actual performance, such as the reference to a given score that tells me which notes to play next *independently* of how and what has been played to the moment. These factors in the shaping of future sounds are more or less distributed processes: They define an identity of the performed piece of music that is distributed among the multiplicity of agents. This is why we call this process one of a distributed identity. And why we would define the quality of the present performance as being defined by the coherence and strength of this flow as a distributed identity. We shall discuss distributed identity as a generator of musical time in detail in Chapter 10.

Putting the three components of performance together, we see *first* that the pairing of body-instrument-body and of gestural space vibration could be conceived as the aspect of dance, and as such, knowing how what is done, expressing beauty. Combing the respective rationales, a *second* pairing, body-instrument-body and the flow of structural unfolding, answering the what and why questions, would then be seen as the balance in a bodily realm, a concept that is akin to what classical Greek aesthetics called "kairos"—the perfect balance in the body's dynamics of presence, and expressing the musician's responsibility. And *third*, the pairing of gesture and structural flow would be understood as a shaping of the body of time, the dynamics within the imaginary time-space that defines our imaginary body of time; we might call this the presence in performance, it answers the questions of knowing how what is performed. Putting these three pairings together, one could then conceive *musical performance as being the balanced dancing of presence*, an existential intensity in three directions of communicative activity.

# Part II

# General Time Concepts

# 2

# Time in Philosophy

**Summary.** This chapter summarizes philosophical perspectives on time. Our choice starts from Plato, includes Chinese aspects, but also deals with modern approaches by Paul Valéry and Maurice Merleau-Ponty, for example. The point of starting with philosophy is that time has always been a deep concern in philosophy. It is fascinating that from early philosophical days, there has been a consciousness of the double nature of time: its linear objective and its self-referential subjective reality.

$-\Sigma-$

## 2.1 Plato

**Summary.** In classical Greek philosophy, time is a double concept: Chronos vs. Kairos. Chronos signifies the given flowing time of the universe, while Kairos is the delicate balance of the presence between past and future. For Plato, time was a parameter of projection of his Platonic ideas, which live outside of temporal categories, in the existential cave of human existence.

$-\Sigma-$

Chronos was a child of the heavenly god Uranus and the earth goddess Gaia. It expresses itself in nature from blossoming to withering, in the lifetime of mankind or the existence of the universe since the Big Bang. Just as we often judge our time, the god Chronos was cruel, hard, tormenting, , see Figure 2.1. According to mythology, he killed his father Uranus to have sole power. In order not to be murdered by his own children, he devoured all his children, with one exception: Zeus. He was born and hidden in the shadows of Rhea. Zeus was thus considered a symbol of the fact that anyone who takes responsibility for his own life can escape Chronos' clutches. This means that everyone should define time as his time and become aware of his own lifetime. Those who tend to let themselves drift, leave decisions to others, assign blame to others, will

 
G. Mazzola et al., *Making Musical Time*, Computational Music Science,
https://doi.org/10.1007/978-3-030-85629-8_2

then become chronically exhausted, chronically overstrained, chronically ill and die. By the way, when Zeus was great, he made Chronos spit out the devoured siblings Hestia, Demeter, Hera, Hades and Poseidon. Zeus married his sister Hera and had four children with her, as well as with his loves, among others, his daughters Helena, Aphrodite, Athena, Iris or also Heracles, Hermes and Donysos.

Fig. 2.1: Chronos TBD (Giovanni Francesco Romanelli, National Museum in Warsaw, a 17th century depiction of Titan Chronos).

Chronos was for Plato (and also for Netwon) the a priori empty container that exists independently of its content, the world's objects. But chronos time was not the fundamental existence, the world of humans was only a projection into the cave of our existence stemming from atemporal ideas. Chronos time is in Plato's philosophy only an attribute of our existence, not a basic ontological entity.

The youngest son Kairos of Zeus is the second of the Greek gods of the time. He represents the right moment in Greek mythology. Kairos is figuratively described as a young man with little wings on his feet, always scurrying around on tiptoes, never standing still, see Figure 2.2.

Nobody knows when he will be where and when he is there, he is already gone. Very striking is his mop of hair and his bald skull, as well as the razor sheep in his hand. Kairos coined the phrase "seizing the opportunity by the hair", which means that you should be prepared for the right moment, then

Fig. 2.2: Kairos TBD (Fourth century BCE marble representation of Kairos by Lysippos from the Museo Archeological in Turin).

you can grab it (grab the hair). If not grabbed in time you can slip off the bald back of your head. Kairos gives time a completely new dimension. With courage and decisiveness, one goes into joyful action. Take a risk, make mistakes, gain experience, take responsibility for yourself, your thoughts and feelings, your own life. The key to happiness lies in Kairos and thus the right moment, the responsible courageous action.

This concept goes way beyond Plato's chronological concept of time. He considered Kairos as a merely rhetorical technique, not as an ontological dimension, he condemned its usage. In modern terms, Kairos is the first conceptualization of presence, the only aspect of time which we actually experience. Refer to our general architecture of artistic presence as described in Section 1.5.

**Impact.** The duality of Chronos vs. Kairos is an early insight into the two ontologies of time: Chronos is, similar to Newtonian time, that line where we are embedded in a fundamentally passive way. For Plato, this is characteristic of the platonic cave, where we subsist as shadows of atemporal platonic ideas. Typically, for Plato, the Kairos time was not an ontological category, he did not understand that the presence in time as described by Kairos is ontologically substantial, not just a rhetorical construction. In the perspective of our book's basic thesis of musical time being a gestural construction, Kairos is a dancing reificton of time, your existence in time unfolds from your balanced dancing presence, that delicate inner substance as embodied from your active gestural existence. Existence in the Latin etymology means *ex sistere*, "to step out" from the simple being that would also be shared by a stone.

– $\Theta$ –

## 2.2 Aristoteles

**Summary.** Aristoteles is the first Greek philosopher who has ever stated an explicit and concise definition of Time. Opposed to Plato, Aristoteles considers time as the amount of motion on a continuum of pre and post (before and after) other than the motion itself. Time for Aristoteles is a global order in which all things get related.

– $\Sigma$ –

Fig. 2.3: Aristoteles, the realist.

Aristoteles, Leibniz, and others have argued that time does not exist independently of the temporal events. This theory is called "reductionism with respect to time" or "relationism with respect to time," since from this perspective, all time talk is reduced to talk about temporal relations among things and events. This means that in this understanding of time, only the events among objects reify time, it cannot exist in an empty space without events and objects. The interesting news here is that time is the result of a construction among events. Not only human events, but any events. This shift from a "divine" Platonic chronos to a secondary category that is derived from events is remarkable and will be in the focus of many philosophical and cognitive approaches. And yes: our main thesis is about the construction of time in music, a radical extension of Aristotelian ideas.

Aristoteles agreed on the ontological status of Kairos with Plato as a merely rhetorical device.

In both, Platonic and Aristotelian time concepts, the ontological status of time is arcane. In particular, in the Aristotelian approach, the temporal order between events is not explained: where, in which layer of ontology does this order appear? In what events and objects?

**Impact.** The interesting impact of Aristoteles' time concept is the introduction of events as substantial components of temporal experience. Despite of Aristoteles' agreement with Plato's (dis)qualification of Kairos as a merely rhetorical category, we now perceive the introduction of an explicitly gestural understanding when basing time on event-related phenomenology.

– $\Theta$ –

## 2.3 Kant

**Summary.** For Kant, and opposed to Newton, time is not an external parameter, but an a priori form of our perception/representation, like space. It is critical here to understand the *conditio humana* of this approach.

– $\Sigma$ –

Immanuel Kant, in his *Kritik der Reinen Vernunft*[1] [65], writes:

> *Die Zeit ist nichts anderes, als die Form des inneren Sinnes, d. i. des Anschauens unserer selbst und unseres inneren Zustandes.*

Which tranlates to:

> *Time is nothing else than the form of the inner sense, i.e., the view of ourselves and our inner state.*

Fig. 2.4: Immanuel Kant argued that time is an a priori inner human sense.

This approach is in sharp contrast to Newton's (and Plato's) idea of time being a "container" of things out there. For Kant, time is a human a priori way

[1] Critique of Pure Reason.

of viewing the perceived empirical reality. He takes the same approach for our concept of space. It is a constituent of our "inner state", which is not empirical, but a *conditio sine qua non* of human world perception. Space is specified as an a priori view of external objetcs, while time is more globally present in external and internal human perception. He concludes:

> *Alle Erscheinungen überhaupt, d. i. alle Gegenstände der Sinne, sind in der Zeit, und stehen notwendigerweise in Verhältnissen der Zeit.*

Which translates to

> *All appearances, i.e., all objects of the senses, are in time and necessarily are in temporal relationships.*

This approach raises a number of questions. To begin with, the Platonic idealism of the divine Chronos idea is transferred into the realm of human a priori. Time is not something out there, but a genuinely human entity. The world out there has no time, but is given a temporal perspective as a human entity.

Moreover, the Aristotelian claim of time being a relationship among events is being altered to the claim that this relationship is not among events, but among their projection into our cognitive stratum.

The big question here is about the genealogy of time in the human a priori. How can we make time? For Kant, this question is not posed, the qualification of an a priori entity eliminates the genealogical approach in Kant's strategy. Time and space are simply there, they don't have to be constructed.

**Impact.** Kant's revolutionary step is the shift of time's ontology from the idealistic Platonic sky into the human ontology. For Kant, time remains a given entity, it is not a human construction. But it is inside human existence, it no longer belongs to a super-human ontology. The shift from Newton's and Plato's external substance, where we are embedded without any possible contribution, to the internal, however a priori, ontology of humans, is the opening of a program (however not enunciated by Kant) that should investigate the mechanism of human time reification. The logical consequence of Kant's shift is in fact this program *if* we acknowledge that our inner a priori of time could be a human process, and not a God-given implantation.

– $\Theta$ –

## 2.4 Valéry

**Summary.** Discussion of Valéry's time construction philosophy in his *Cahiers* [161] and Berkay Ustun's paper [159].

– $\Sigma$ –

Fig. 2.5: Paul Valéry was one of the first thinkers who maintained that human time is a construction.

Valéry, unlike Plato and Newton, does not argue that time is an a priori empty container to be filled (or not) with events or experiences. Going in the same direction as Kant, but delving deeper in the human conscience as a principal element of the notion of time, Valéry considers "time as being production" [161]. Recall that for Kant time is construction—however, the German philosopher does not make explicit how the genealogy of this construction developed until reaching the form in which we experience it. Instead this construction, akin to Plato's idea of time as an independent entity existing outside of our experience, is also imposed on us a priori, and we apparently have no ways to escape the constraints of this time that the humans ourselves created somewhere in the past.

By affirming that time is production, Valéry highlights the vital importance of human conscience in the production of time experiences and how different states of mind, such as attention or surprise, might affect the manner in which the time is modeled. "Time and consciousness are intimately related"[2] [161, Vol. II, p.224] points out to the underlying function of the individual's conscience in producing his/her own time judgments as a "residual mental phenomenon"[3] [161, Vol. II, p.181] coming from the different states of mind and perception that inflect what we call time.

But if time, to Valéry, is essentially unmetrical and dependent on the perspective of the subject that produces it—here a subject taken into all its parts, mind and body with all its components—how is time then perceived? Through the processual work *Cahiers* [161], the writer states that the construction of

[2] Le temps et la conscience sont intimes.

[3] Le temps est un phénomène mental résiduel.

time finds its underpinnings in the perceptions of difference: "time is the generic name in all the perceptions of difference."[4] [161, Vol. II, p.172] (or "time is a collection of different portions; each one containing different things that coexist"[5] [161, Vol. II, p.125]. Hence, time is the mental phenomenon that is produced as a residue of perceptions of difference. This awareness of difference is based on a duality of things that are recognizable as a single entity but presenting, nevertheless, a change in its existence [161, Vol. III, p.538]:

> Le temps – en somme – est ce qui peut être construit ave n'importe quelle diversité d'éléments sous certaines conditions comme par exemple d'être à une dimension – C'est la substitution de n'importe quoi à n'importe quoi avec distinction ou asymétrie des deux termes ou quelque chose qui les distingue quels qu'ils soient (fussent-ils identiques).

Which we translate to

> Time—in sum—is what can be constructed with any diversity of elements under certain conditions such as for example being in one dimension—it is the substitution of anything for anything with distinction or asymmetry of the two terms or anything that distinguishes them whatever they are (if they were identical).

This one-dimensionality of the changes in time relates to the Aristotelian idea of time as being the amount of motion spent in a single axis without, however, considering the motion itself. The main point is here an unbalance of objects that are experienced as being the same but change their positions in this one-dimensional continuum, happening in succession and inflecting the perspective of the observer [161, Vol. III, p.72]: "Time is a valuation, a function of two or n states that declares them as successive."[6] This thought is also close to McTaggert's "A series"—some events take their positions as present, and then other events have their positions constantly passing from the future to the present and from the present to the past, accounting for an everlasting change in the events' placement without, however, altering their order, see also Section 2.6.

Valéry goes beyond the formal understandings of time arguing for a time that is body-dependent, embodied, poietic (and also aesthesic). To Valéry, the body is not only seen as in traditional cognitive entity, as a brain (acting as a master) with its mental representations, governing a body (playing the role of a vassal). The body is taken in its full ramifications, with the muscles and the limbs—the gestures, in short—acting as important building blocks not only of human experiences in general, but also of the form in which we perceive

[4] Le temps est le nom générique dans toutes les perceptions de différence.

[5] Le temps est une collection de portions différentes; chacune contient des choses différentes coexistentes.

[6] Le temps est une appréciation, une fonction de 2 ou n états qui les déclare successifs.

and produce time. Thus, time cannot be generalized (as Kantian time, that works for all human beings, as an all-encompassing form of sensibility) but as stated by Berkay Ustun [159, p. 6], "instead of assuming a universal and absolute time to find it applicable in individually localized set-ups later, it (Valéry's understanding of time) always starts from the closure of an observably local functional cycle that is capable of undergoing transformations and modulations."

Valéry's thoughts also account for the multiplicity of possibilities for the construction of time in music, by considering the plurality of 'times' that can arise from different contexts or local systems—such as different music compositions or different interpretations of the same piece—through the actions of the musicians' minds and gestures. The fixity of time is rejected in favor of a time that is ready to be newly invented, produced and recreated, "There is no such thing as time in general. There are only the times belonging to systems." [161, Vol. IV, p.XXX]

**Impact.** Valéry's approach to time is a major step from Kant's abstract time a priori to a proper human *construction*. The non-existence of "time in general" implies that the individual human becomes the generator of time. Valéry introduces two new thoughts: time is a human construction, and it is an individual construction. But a third new perspective is that Valéry also delegates the responsibility for this temporal construction to a mechanism of the *totality of human embodiment*. To our knowledge, this is in the history of philosophical ideas the first moment, where this radical perspective of time's ontology is set forth. What is still missing in this approach is an explicit description of how time is derived from the embodied construction.

– $\Theta$ –

## 2.5 Chinese Philosophical Time Concepts

**Summary.** In ancient Chinese philosophy, as transmitted in Confucian and Taoist traditions, a remarkable absence of the theme of time is manifest. The French philosopher François Jullien in [63] (Figure 2.6) discusses this difference without classifying it as a deficiency, but as a radically different understanding of time, which is not an abstraction from events and their movement, but it is engrained in these processes.

– $\Sigma$ –

Jullien analyzes the difference between European and Chinese understanding of temporality. He concludes that the principle of Yin and Yang of an incessant processuality between these two polar forces implies that "time" is not a concept and is instead replaced by "process". In Chinese language, the verb "to be" is inexistent, it only appears in the function of a copula. Chinese do not oppose the a-temporal to the temporal. Jullien refers to Marcel Granet's

Fig. 2.6: François Jullien discusses Chinese time philosophy.

fundamental work on Chinese thinking [51] as being the first (non-Chinese) instance of having recognized this radical difference to European approaches to time, see especially Chapter 1: Time and Space of the Second Book. His conclusion coincides with Granet's insight, namely that China was led to think not of "time", but of *process*.

Fig. 2.7: Marcel Granet understood that China was led to think not of "time", but of *process*.

And this process was always a process of events involving objects of a given (although often complex) reality. Granet states that: "Aucun philosophe chinois n'a voulu voir dans le Temps un paramètre."[7] This difference is significant since the abstract time concept is replaced by a temporality that is induced by or derived from processes, i.e., in particular, a temporality that is a secondary effect being constructed from processual interaction. This is relevant to our own approach to musical time, see Chapter 11.

In Chinese language, verbs cannot be conjugated, they have no temporal specification in their grammatical structure. Jullien refers to Saint Augustin as

[7] No Chinese philospher wanted to view time as a parameter.

the first among European philosophers who has understood that our temporal conceptualization is fundamentally based upon our language. The tripartition past, present, and future is a language principle and our thinking of time is derived from this linguistic base. Jullien recognizes that Saint Augustin's analysis of time was precisely driven by his Latin language.

Jullien analyzes the Chinese shape of temporally sensitive concepts and recognizes two basic instances: the moment or occasion (shi, 时) and the duration (jiu, 久). However, these concepts are not included in an abstract comprising concept of time. The first, moment, is strongly related to seasons (si shi, 四时) as moments in the changing biological state of their environment. The second, duration, is not abstract, but always thought of as being the duration of something. The abstraction from this object that endures is not made in Chinese thinking.

At the end of the 19th century, the European concept of time was translated to Chinese as "between moments" (shi jian, 时间). The Chinese duration is a concept that refers to the transition from moment to moment. This transitional reality is a kind of deepened moment, not its negation. This seems interesting in our own development of a deepened understanding of the making/construction of time in music: Time is not an abstraction from events and their movement, but it should be engrained in these realities.

Jullien asks a critical question about the European approach to time: Could it be that the European philosophers have never really investigated the very concept of time and instead only operated their temporal analyses to describe other items that use time as a regulatory parameter? This observation is important for our own approach to time, since we shall not just apply time to construe the musical concept space, but we shall most importantly investigate its very definition.

**Impact.** The Chinese approach to time is deeply different from the European one. There is no immanent concept of time, it is only a side effect of a processuality among events. This approach opens up a constructive understanding of time as a phenomenon that is implicit in the interaction of processes. Time is not a Chinese concept. We view this perspective as a radically different one, which focuses on events, on gestures and not on their potential parametrization by a time line. This complies with our basic thesis that human time conceptions are derived from gestural interactions, and not from their parametrization by a specific time dimension.

– $\Theta$ –

## 2.6 John M. E. McTaggart: The Unreality of Time

This section refers to McTaggart's book *The Unreality of Time* [109]. He distinguished two ordering of temporal events. First, time can be ordered by the relation of *being earlier than*. McTaggart calls the ordering generated by this relation the B-series. The second ordering is generated by specifying a moment within the B-series as the *moment of presence*. McTaggart calls this ordering the A-series. Following McTaggart, for time to be a reality both series must coexist.

Fig. 2.8: McTaggart's argumentation against the existence of time is that reality of time means that change takes place.

John McTaggart's argumentation against the existence of time is that reality of time means that change takes place. And that this must be reified within the A-series as a succession of changing moments of presence. But he argues that the A-series does not exist since nothing is really ever present, past or future. He argues that being present, being past, and being future are incompatible specifications. And the A-series would ask for such contradictory specifications of its elements.

This argumentation is in fact a corollary of McTaggart's concept of the ontology of presence. Presence does not subsist, it disappears *ontologically* when it is over. For our present study this position matters as far as it evokes the fundamental question of the time's ontological status, and more precisely the status of presence.

## 2.7 Edmund Husserl: Vorlesungen zur Phänomenologie des inneren Zeitbewusstseins

Husserl in [58] analyzes our phenomenological perception of time as an activity that transcends the mere sensual input. The perception of a melody is generating a temporally ordered succession and relation among the melody's tones. This performance is added to the sensual input as a temporal dimension.

Fig. 2.9: Husserl analyzes our phenomenological perception of time as an activity that transcends the mere sensual input.

## 2.8 Henri Bergson: Mathematical vs. Pure Time

Bergson proposes two forms of time: pure and mathematical time. The former shares a real duration and differs from the latter that is perceived via distinct units. Pure time or lived consciousness is continuous and indivisible. Mathematical time can be divided into discrete units, seconds etc.; these units do not comprise the flow of pure time. Bergson also argued that time as measured by the hands of a clock is a spatial abstraction. But lived consciousness is a bundle of space-time continuity. It is indivisible. The flow of real time can only be known by intuition [14, pp. 75-101].

Real (pure) time which is not a series of moments, a succession of acts of direct presence in the concrete experience. To this end—to know the real experience of duration—one must be part of the awareness of the present entity.

According to Bergson, pure time cannot be analyzed mathematically. Any measurement of time presupposes a breaking or disruption of time. Real duration can only be experienced by intuition. For our approach to time Bergson's

Fig. 2.10: Bergson proposes two forms of time: pure and mathematical time.

distinction of time concepts is interesting as it stresses the objective mathematical time line as qualitatively different from the subjective and intuitive reality of time in human perception. It is however not clear so far, how the coexistence of these two concepts would be reconciled as two perspectives of a unified more complete concept of time.

Fig. 2.11: Jean Wahl's point of view is similar to the Chinese understanding of time, namely as a construction that is implied and derived from the events' collaboration in succession.

## 2.9 Jean Wahl: Quality of Events

Jean Wahl's "Si le temps est qualité, il est qualité de ces événements qui se précèdent ou succèdent ou sont contemporains les uns des autres."[8] [162, p. 308]. This point of view is similar to the Chinese understanding of time, namely as a construction that is implied and derived from the events' collaboration in succession.

**Impact.** McTaggert, Husserl, Bergson, and Wahl share the doubt concerning the uniqueness of a time concept, they introduce special aspects of the original Greek time duality as driven by their specific philosophical approaches. Nevertheless, their contributions are not as far-reaching as Valéry's revolutionary insights.

– $\Theta$ –

[8] If time is a quality, it is the quality of these events that precede or succeed each other or are simultaneous.

## 2.10 Maurice Merleau-Ponty: La Phénoménologie de la Perception

We refer to Maurice Merleau-Ponty's idea that time as a construction of consciousness (*La Phénoménologie de la Perception, Partie III, Chapitre 2,* [107], English translation [108]).

Fig. 2.12: Maurice Merleau-Ponty maintained the idea that time as a construction of consciousness.

The French philosopher Maurice Merleau-Ponty privileges the body as being the center of all human experiences. Time, hence, depends not only on our imposition of certain qualities that define its perceived passage, but also on how this time is embodied by us, how it is lived through our bodily interactions and gestures. Merleau-Ponty was a phenomenological philosopher, and his understanding of time arises from the disagreement with the views of two other phenomenological[9] philosophers regarding this matter: Bergson and Husserl. This disagreement implies a critique of both the empiricist (Bergson, see also Section 2.8) and intellectualist (Husserl, see also Section 2.7) phenomenologies and the conceptions of time they entrain.

*Empiricism* postulates an objective view of the world, entailing a conceptualization of it as being born in the interactions of the subject with its medium by means of sensations: the role of the subject is then limited to taking sensational cues from the world, whose gathering would eventually lead to experiences. For empiricists, the subject is derived from the object. The subject finds him/herself inserted the world, and seeks to understand it without resorting to any advanced knowledge besides the very sensations the world offers him/her. The world is out there, as a background over which the subject's consciousness has merely to be accommodated. The world itself is only a qualitative category

[9] Phenomenology is a philosophical approach, which focuses on the phenomena as opposed to hidden entities. This means that what is a phenomenon of our experience is the starting point of this perspective of thought.

without a valid time concept—the Bergsonian conceptualization of the timeless world—and time exists only because humans, as subjects, are inserted in it, hence needing to accommodate their consciousness in it. This accommodation takes place by means of the introduction of the instant into consciousness, that will ultimately establish the objective succession of instants—or presents—that characterize the spatialized time, the time that is not a continuum, not a flow, but a procession of the infinitesimal, a procession of instants, that define time in an empiricist view.

*Intellectualism* does not limit the role of the subject to the passiveness of a receptacle of sensations. Intellectualism denies the idea of the empiric subject, whose only active part in the role of being in the world is to create abstractions of it—the experiences—a task that can be accomplished without any advanced knowledge of the world, since the minimal effort necessary to accommodate one's consciousness in it is also enough for its comprehension. Intellectualism introduces the concept of judgment, an analytical reflection on the experienced, seeking to adapt an intelligible structure to the world, overcoming the limitations of empiricism by means of the imposition of a 'conceptual supplement', a tool serving to correct the errors and imprecisions of perception. In the domain of time, this entails the Husserlian view of a 'time of inner time', a time that belongs to the subject, that is inextricably linked to him. This reflective time is constituted by the subject, and leads to a 'duplicity of times'—the empiric time of the world, coming from the subject's perception (the empirical observation is not abandoned); and the inner flow of time of the subject, the conceptual supplement that mediates the relation of the inner time with the exterior time.

These two aforementioned conceptualizations of time, in Merleau-Ponty's perspective, account for the very failure of the phenomenon's comprehension. Since they place an abstract consciousness—somewhat similar to the Cartesian *cogito*—at the center of the phenomenon, they take for granted the role of the subjects and their concrete position in relation to the events perceived by them. This does not account for a negation of consciousness, but for a necessity of understanding the consciousness as interacting with its medium and having a much greater role in temporality than just gathering cues from the environment or measuring the time of the world against a conceptual time, which remains inherently imprisoned in a consciousness that is detached from the world with which this consciousness is interacting.

Merleau-Ponty advocates for an embodied consciousness, in incessant communication with the subject's body. There cannot be, in Merleau-Ponty's vision, a consciousness without a body, for the subject exists in the world precisely because of the body that allows itself and the consciousness to interact with their environment. The boundaries between mind and body become blurred in favor of an idea of communication. Data are not anymore sealed in the consciousness, because incommunicable mental representations lose their meaning and their connection to the world that made them possible. Merleau-Ponty argues that time has to be understood from the perspective of an embodied consciousness inhabiting a world, not only as pure dimension of the mind:

"let us no longer say that time is a 'datum of consciousness', let us be more precise and say that consciousness deploys or constitutes time" [108, p. 481].

Merleau-Ponty's phenomenology of embodiment also opposes his conception of time to that of Kant, whose understanding of time placed the phenomenon as one mere category of internal *a priori* perception. Kant, in Mearleau-Ponty's view, thinks of time as being solely this 'datum of consciousness', a consciousness that constructs temporality disregarding the external world or transforming it through mental operations that cause the world, once having been transformed into mental representations, to lose its meaning. Merleau-Ponty (in agreement with Heidegger on the other hand, argues that humans do not create temporality, but they temporalize the world through the interactions of their situated embodied consciousness. If time were merely a category of perception, as Kantian time is, then the experience of time would be homogenized, a perspective of time that even implies, for Merleau-Ponty, a negation of time altogether: "time as the immanent object of a consciousness is time brought down to one uniform level, in other words it is no longer time at all" [108, p. 482]. This conclusion rests on the implicit assumption that time is a fundamentally individual experience, and therefore, a uniform time concept is a failure.

Merleau-Ponty's thought displays some similarities with Valéry's conceptions, especially regarding the subjectivity of time. Remember that for Valéry time is a production (see Section 2.4), derived from the subject's individuality, and always undertaken afresh. Merleau-Ponty, however, goes deeper in the quest for an embodied construction of meaning, placing the body at the center of phenomena. Perception is no longer imprisoned in the *cogito*, but residing in the body, for there can be no interaction with a world without a body which inhabits it. Here, the cartesian *res extensa* of the body is replaced by a substantial role of the body in the constitution of our very existence. It is no longer "cogito ergo sum"[10], but, as we may say, "corporeo ergo sum"[11], and then the experience follows from the embodiment. The following examples, very illustrative of Merleau-Ponty's considerations on perception, reflect the ideas of his philosophy of embodiment. "The color, before being seen, gives itself away through the experience of a certain bodily attitude appropriate only to that color and precisely indicative of it" [108, p. 244] or "within a musical note a 'micromelody' can be picked out and the interval heard is merely the final patterning of a certain tension felt throughout the body" [108, p. 245]. Perhaps what Merleau-Ponty is trying to illustrate is that we sense things, with the whole constitution of our bodies, before we can make sense of things.

We have seen in these examples how Merleau-Ponty relates the perceptual experience to bodily responses that join their signification, complete their meaning, implying the whole constitution of the body in the phenomenon of perception. Merleau-Ponty's argument for the indissociable co-dependence of

[10] I think therefore I am.
[11] I embody therefore I am.

mind and body is made clear several times throughout his text: "the actual existence of my body is indispensable to that of my consciousness" [108, p. 501]. This new principle is derived not only from the critique of the phenomenologists, but also from Merleau-Ponty's summary of all of his investigations about body and consciousness. If time is no longer a mere category of perception, but something that is constituted by consciousness, and the body is indispensable if the subject is to be invested with consciousness, then the body has a pivotal position on the constitution of time. "Time is, therefore, not a real process, not an actual succession that I am content to record. It arises from my relation to things".[108, p. 478] But how does this time emerge from my relation to things?

Merleau-Ponty's approach to describe time is a quintessential demonstration of the phenomenological approach of doing philosophy: he does not seek for an explanation, but rather looks for a description of what time is. First, Merlau-Ponty concurs with Bergson's timelessness of the world. Our perspectives as humans will then constitute temporality. Like the notion of movement in physics, the necessity of a referential point that can attest the movement in relation to it, time depends of a consciousness ready to deploy it. For Merleau-Ponty that is the cause of the confusion of treating time as a variable of nature or a form separable from its matter. Just like in physics: the movement ceases to exist in the absence of a referential. Likewise, the adoption of any referential besides the individual embodied consciousness destroys the real nature of time.

The privileged position for the constitution for the process of temporalization is the present. "Is in the present that being and consciousness coincide" [108, p. 492], more specifically in what Merleau-Ponty calls 'field of presence'—the position where the subject makes contact with time. This field of presence conflicts with the notion of spatialization of time, for that time is not experienced in the succession of the infinitesimal. The present ceases to be when alterations force it to move to the past, what recalls Valéry's postulation that the notion of time is based on the perception of difference. An event ceases to be present when some alteration in my field of presence forces my presence to move into the past. But for Merleau-Ponty, when an event moves to the past it does not cease to be. One can still rejoin it, by re-establishing the configuration that made it possible, or one can effectively move into the future, by letting one's present sink into the past as the layer of time thickens and what was once my past now gets progressively more distant. Time, hence, is not a line constituted of the succession of infinitesimal presents—the spatialized time—but a network of intentionalities [108, p. 484]. Subjects have the power to effect the passage of one present to another [108, p. 487].

It is noteworthy that the very notion of event, inexistent in the conservational nature of the world, becomes an important tool of our consciousness in the constitution of time. Bergson's notion of a continuous time, whose transitions from present to past are imperceptible, denies time in Merleau-Ponty's view, since time needs change. And the order of events, once it has been established by the succession of intentionalities that constitutes time, does not destroy itself. We can think of McTaggert's B series—the temporal order of the past remains preserved (see Section 2.6).

For Merleau-Ponty, the process of objectivizing time (effect its measurement by means of clocks) is not enough to account for its deep relation with our consciousness, as real time means more than just awareness of its passage. Time becomes the very medium where consciousness and being are deployed. The relationship between time and the subject is deeper than just recording its flow: time and the subject are inextricably united, one needs the other in order to exist, their relation is conceived from within, the subject needs to be temporal, as "virtue of an inner necessity" [108, p. 476]. "Time is a dimension of our being" [108, p. 483], a being that is integrated with the world our body occupies, where my gestures define the meaning of our interactions, indissolubly with our consciousness. For Merleau-Ponty, we are time and time is us, time is in us, and Heidegger's statement, also quoted by Merleau-Ponty at the beginning of his chapter on Temporality, is very appropriate to summarize much of his ideas about time: "The sense/meaning of being is temporality"[12].

**Impact.** Merleau-Ponty inherits from Valéry the individual character of time, its construction and its embodied reification. But he goes deeper into the existential role of time: For him, time is characteristic for our existence, its identification. As such, it is a thoroughly individual category, and in particular, an uniformization of the time concept negates time from the beginning. Merleau-Ponty performs his own a priorism in that his time philosophy connects time to individual existence as an identifier: time is existence, and more precisely: embodied existence. In opposition to Valéry, who maintained that someone who tried to reduce thinking to neuroscience is an "imbécile" (an idiot), Merleau-Ponty attempted such a reduction. However, much as Valéry, Merleau-Ponty does not describe *how* his embodied time concept creates time, he gives us a qualitative theory, whereas the concrete gestural mechanism that leads to time remains arcane.

– $\Theta$ –

[12] "Der Sinn des Daseins ist die Zeitlichkeit." (Heidegger, Sinn und Zeit, p. 331)

# 3

# Time in Physics

**Summary.** This chapter is a summary of some of the dominant temporal conceptualizations in physics. We include Newton's divine time, Einstein's geometric time integration in four-space, Quantum Mechanics, Hawking's imaginary time, including our own perspective upon complex time, and Smolin's critique of physical time concepts that disregard of the qualia quality of time.

$-\Sigma-$

## 3.1 Newton's Divine Time

**Summary.** The Newtonian space-time frame as an a priori setup for physics is introduced.

$-\Sigma-$

Newton's first documented discussion of time and space was written in the manuscript fragment beginning with "De gravitatione et aequipondio fluidorum et solidorum ...", written around a decade before Newton's famous *Principia* [119].

In his *Principia* (Philosophiae Naturalis Principia Mathematica) [119], Newton states (taken from the English translation by Andrew Motte):

> Absolute, true and mathematical time, of itself, and from its own nature flows equably without regard to anything external, and by another name is called duration: relative, apparent and common time, is some sensible and external (whether accurate or unequable) measure of duration by the means of motion, which is commonly used instead of true time...

This claim in particular means to Newton that humans cannot perceive absolute time, it is a mathematically abstract reality. And it is unique, absolute time is a singular divine background structure that is given a priori and



G. Mazzola et al., *Making Musical Time*, Computational Music Science,
https://doi.org/10.1007/978-3-030-85629-8_3

Fig. 3.1: For Newton, humans cannot perceive absolute time, it is a mathematically abstract reality.

does therefore not involve physical events. They are embedded in this absolute, immutable, and non-empirical frame.

Newton's space-time concepts have been criticized by the point of view that all inertial frames of reference are equivalent and related by Galilei's relativity theory (which nevertheless keeps time invariant). Already Gottfried Leibniz, whose approach was criticized by Newton in [119], maintained that without bodies and events, space and time are meaningless. In the 19th century, Newton's position was questioned by physicist Ernst Mach, who stated that space-time is a reference to the "whole universe", and not to a transcendental ontology.

For our concerns regarding musical time, Newton's position is the extreme of a European approach to time as an autonomous ontological dimension that is given as an immutable frame, and where all events are embedded a priori.

The difference to Kant's time concept is that his a priori of time is an internal human perspective, while for Newton, this is an external objective reality that humans cannot perceive.

## 3.2 Relativity Theory

**Summary.** We discuss Albert Einstein's revolution form Newton's singular time to a time manifold given by the variety of reference systems and the connecting Lorentz transformation. Kurt Goedel's model of an Einsteinian universe with a circular time.

– $\Sigma$ –

Einstein's special relativity performs two radical changes from Newton's approach to time: First, he views time and space as connected within a four-dimensional space of "space-time". Second, the transformation between different inertial frames of reference is described by the Lorentz transformation that connects time transformation with the relative velocity $v$ of two such frames. The typical formula of this transformation, movement in $x$ direction, is

$$t' = \gamma\left(t - \frac{v}{c^2}x\right) \tag{3.1}$$

$$x' = \gamma(x - vt) \tag{3.2}$$

$$y' = y \tag{3.3}$$

$$z' = z \tag{3.4}$$

where $x, y, z$ are the space coordinates in the unprimed frame, and the $x', y', z'$ are the space coordinates in the primed frame. The unprimed time coordinate is $t$, and the primed time coordinate is $t'$. The light velocity is $c$, and the Lorentz factor is $\gamma = \sqrt{1 - v^2/c^2}$. The primed time now depends on the relative velocity $v$.

Einstein's revolution however does not change radically the ontological status of time, it simply connects time to space and then allows for a 4-dimensional transformation of space-time. But the ontology of this space-time remains an external frame, much in the spirit of Newton. It relates to the inertial frames of reference, but it is nevertheless detached from the objects and events in such frames. This ontological situation is preserved for time dilation in general relativity, which is cause by gravitational fields; however, now the presence of gravitating masses influences the time's values. But it does so in a Newtonian transcendence, no human action is involved, time is simply there as a given world parameter.

It is known that the famous logician Kurt Gödel, who spent the final part of his life together with Einstein at the Institute for Advanced Study in Princeton, presented a periodic solution of Einstein's equations of general relativity theory [44]. His conclusion on the physical concept of time shows the possibility of a circular time trajectory, i.e., future and past become relative because on a circle, the linear ordering is only locally defined. More radically speaking, Gödel concluded that time is an illusion [170].

## 3.3 Quantum Mechanics (QM)

**Summary.** In QM, time is not an observable, it is abstract and external, similar to Newton's "divine" time.

– $\Sigma$ –

The double reality in QM between the Hilbert space H, plus observables (linear operators on H) on one side, and their projection, via eigenvalues of observables in experimental measurements, on the other, is discussed with a focus on the remarkable fact that time in QM is not an observable. In QM, time is not an observable, there is no experimental access to time, it is, similar

Fig. 3.2: Albert Einstein and his friend Kurt Gödel at the Institute for Advanced Study in Princeton.

to Newton's "divine" time, a given 'absolute' variable, for example used in the description of the wave function $\psi : \mathbb{R}^4 \to \mathbb{C}$ of a single particle (three real space coordinates, one time coordinate). The *Hamilton operator* $H$, an expression for the observable's energy of an observable $\psi$, describes the temporal behavior of the observable by the *Schrödinger equation* $i\frac{d}{dt}\psi(t) = H\psi(t)$. Also, in contrast to Relativity Theory, QM does not integrate time and space.

Fig. 3.3: Steven Hawking introduced the idea of complex time in cosmology.

## 3.4 Complex Time

**Summary.** We discuss the complex time concept (complex numbers) in physics, introduced by Stephen Hawking and Izak Bars to solve cosmological singularity problems of the Big Bang.

– $\Sigma$ –

In physics, more precisely: cosmology, the real-valued time $t$ (the time line being the set $\mathbb{R}$ of real numbers), has been extended to complex time $c = t+i.s$, element of the set $\mathbb{C} = \mathbb{R} \oplus i.\mathbb{R}$ of complex numbers. One reason for this extension, as set forth by Stephen Hawking, was the better understanding of the original singularity of the Big Bang. In this model, the imaginary component of time was non-vanishing from the very beginning (in terms of real time) of this universe's existence.

Independently of cosmology, complex time can be used to understand Special Relativity Theory as follows. Recall from Section 3.2 that the Lorentz transformation yields these primed coordinates for $x', y', z'$ for a movement in the direction of the $x$ axis:

$$t' = \gamma\left(t - \frac{v}{c^2}x\right) \quad (3.5)$$

$$x' = \gamma(x - vt) \quad (3.6)$$

$$y' = y \quad (3.7)$$

$$z' = z \quad (3.8)$$

where $x, y, z$ are the space coordinates in the unprimed frame, and the $x', y', z'$ are the space coordinates in the primed frame. The unprimed time coordinate is $t$, and the primed time coordinate is $t'$. The light velocity is $c$, and the Lorentz factor is $\gamma = \sqrt{1 - v^2/c^2}$. The primed time now depends on the relative velocity $v$.

We may forget about the unchanged coordinates $y, z$ in what follows. The Lorentz factor is $\gamma = \sqrt{1 - v^2/c^2}$. All these coordinates are real numbers. One rewrites these equations with $ct, ct'$ instead of the original time coordinates. This yields

$$ct' = \frac{1}{sin(\alpha)}\left(ct - cos(\alpha)x\right) \quad (3.9)$$

$$x' = \frac{1}{sin(\alpha)}(x - vt) = \frac{1}{sin(\alpha)}(x - cos(\alpha)ct) \quad (3.10)$$

with $cos(\alpha) = \frac{v}{c}$ being the cosine of the angle $\alpha$ in the rectangular triangle defined by the vector $\vec{c}t$ and the vector $vt$ on the x-axis in Figure 3.4. The Lorentz factor then is $\frac{1}{sin(\alpha)}$.

This gives us the geometric interpretation of this transformation in complex number as follows: We view the number $ct$ as the length of the vector $\vec{c}t$,

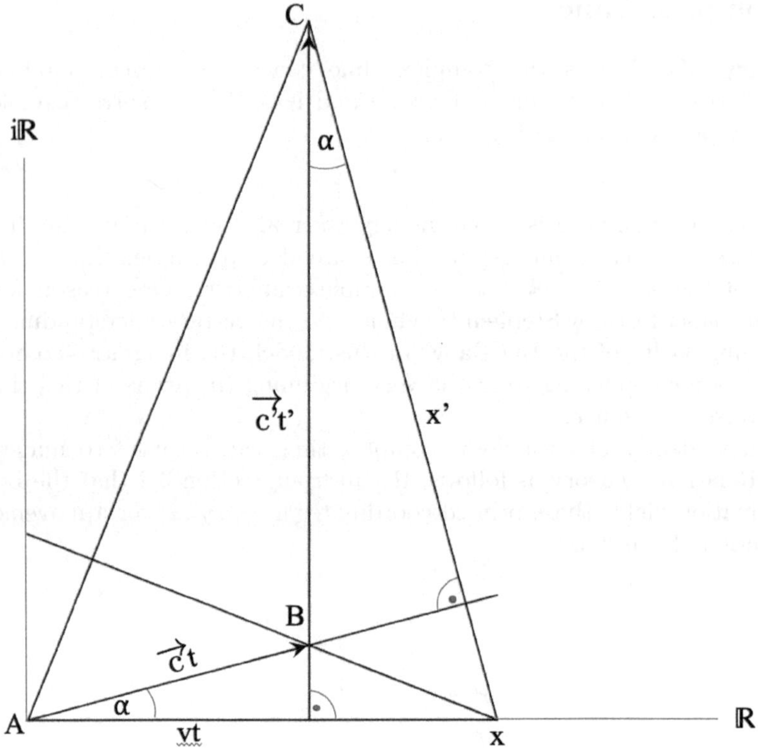


Fig. 3.4: The geometric representation of the Lorentz transformation.

perhaps viewed as the tensor product $\mathbb{R} \otimes \mathbb{C}$ of the $x$-axis with complex time. The $x'$ coordinate is the side opposite to vertex $A$, this follows from formula 3.9. The time component $ct'$ is the negative of the length of vector $\vec{c}'t' = \overrightarrow{BC}$. The velocity vector $\vec{c}'$ has again length $c$, but looks in another direction.

## 3.5 A Physical Interpretation

This geometric representation suggests that the primed coordinates $x', ct'$ are some kind of symmetric versions of the unprimed $x, ct$ along the hight of the big triangle through $B$ and $x$. We want to develop a more physical interpretation of this geometric fact. To begin with, we view this two-dimensional display as being set within the complex numbers. The unprimed reference system has its origin $A$ in the zero of $\mathbb{C}$. The velocity of the primed reference is drawn as a complex vector $\vec{c}$. Its length is $c$, the light velocity, and the position of the arrow head of $\vec{c}t$ is the position of the primed origin (point $B$).

This is a remarkable hypothesis: The primed origin always moves at light speed, but its real projection (onto the first coordinate) has the perceived velocity $v$. In other words, all movements are in light speed, but only in the complex space, the real projection of velocity is what we perceive. For example, horizontal movement in complex space is the photon's movement.

The Lorentz transformation describes the primed reference as a rotated and shifted copy of $\mathbb{C}$. Which would be the new x-axis? Here is the construction.

- First, the new x-axis must be such that the projection of the point $B$, the moving origin, onto this axis shows the primed origin $B$ as resting. This means that the projection onto this new axis must be orthogonal to the vector $\vec{c}t$. This gives us the direction of the primed x-axis.
- Second, we want the point $x$ to be on this primed axis, too. This defines the primed x-axis through $x$ and $C$.
- Third, the primed line segment ending at $x$ must project onto the line segment between the head of $vt$ and $x$. This segment therefore must start at $C$ and end at $x$.
- Fourth, in the primed reference, the movement from the primed origin $B$ must go in the negative velocity $v$ back to the unprimed origin. We have that velocity vector and the time $t'$ that is needed to touch the primed axis $x'$. Its velocity in $\mathbb{C}$ is $c$ again. The distance is $\vec{c}\,t' = \overrightarrow{BC}$, this defines the primed time coordinate $t'$.

This looks like a more physical interpretation of the geometry of this Lorentz triangle. We do not claim that this is a physically important description, but it shows that introducing complex time can open up unseen perspectives, such as the thesis that in (complex) realty, every object would move at light's speed!

## 3.6 Leo Smolin's "Time Reborn"

Leo Smolin recently published a book "Time Reborn" [148] where he defends time as a non-fiction (contradicting Einstein), and where he discusses the problem of consciousness as a fundamental qualia where time becomes a core characteristic. He stresses the difference between the typical physical description of neuronal processes and the ontology (what really is, exists) of consciousness. The physical reality and the psychological, the description and the essence of being, are Smolin's "big questions". In his final chapter, he stresses the central role of time in human consciousness:

> One further aspect of consciousness is the fact that it takes place in time. Indeed, when I assert that it is always some time in the world, I am extrapolating from the fact that my experiences of the world always take place in time. But what do I mean by my experiences? I can speak about them scientifically as instances of recordings of information. To speak so, I need not mention consciousness or qualia. But this may be an evasion, because these experiences have aspects that are consciousness

of qualia. So my conviction that what is real is real *in the present moment* is related to my conviction that qualia are real.

Fig. 3.5: Leo Smolin stresses the difference between the typical physical description of neuronal processes and the ontology (what really is, exists) of consciousness.

We argue that the construction of time is a core substance characterizing this difference in that time is not an illusion, but a fundamental construction of our consciousness. To be clear: a construction is not an illusion, but a fundamental motor of consciousness. Our approach in this book deals with the description and investigation of this construction mechanism. This is the main statement and answer to Smolin's perspective.

**Impact.** The physical contribution to he concept of time has always been one that does not deal with construction or even the idea of a construction of time by human ontology. In his special relativity, Einstein just connected time to space and conceived a multiplicity of such space-times, but without any constructive aspect. A constructive component was added in the general relativity, but only as a material component of gravitational masses to define the 4-dimensional space-time variety. This type of construction is not one that defines time conceptually, it only helps build the manifold's geometric parameters, it is, ontologically speaking, as divine as Newton's time. In QM, they return to Newton's original time line, it is now explicitly non-observable, it lives outside of the experimentally accessible reality. In the complex approach to time, no constructivist aspect is added, it is just a mathematical trick, typically, "Wick rotation" is the physical name of multiplying by the imaginary unit. Only Leo Smolin's approach to the qualia ontology of time (and, BTW, also Roger Penrose's thoughts about consciousness in physics) in human consciousness opens up a window to the question of how time could be a constructive entity.

The impact of this situation in physics is that the quasi-objective reality of physics is missing a fundamental understanding of the nature of time. Penrose and Smolin are by no means exotic physicists, the are only the first to realize that physics is more than a bunch of objects and experiments floating around in a divine space-time theater. And it is not clear, whether the conflict of relativity theory and QM could be related to the present deficient time concept in physics altogether.

– $\Theta$ –

# 4

# Time in Cultures, History and Present

**Summary.** This chapter thematizes the general cultural shape of perception and construction of temporal categories. Our approach is not a comprising and in-depth discussion, but a series of hints to the complex phenomenon of cultural determinants of understanding time.

The concept of a culture needs to be made precise here. "Culture" is derived from the action of cultivating something. Which means to perform a set of activities, but also to perform them in the framework of a conceptual architecture, culture is practice and theory, also coined as "praxis" in contemporary anthropology. This being, a culture is not a static entity, it is dynamic in space and time, it evolves in complex spatio-temporal coordinates. In our approach, we focus on the origins of cultures, not on their evolution in a global context, which would be a much more complex topic.

*– Σ –*

## 4.1 Time in Languages

**Summary.** Different languages have different grammatical and conceptual structures for expressing time. We compare some language families (e.g., past tense is inexistent in Chinese).

*– Σ –*

This section presents some deep and delicate relations between language and time concepts in cultures. We may first cite Benjamin Lee Whorf's famous treatise *Language, Thought and Reality* [165], where he discusses different languages, especially the North-American Hopi language, and sets forth the idea that language is a strong creator of realities, not only their faithful reflection. He writes [165, Ch. 83]: "Thus one of the important steps for Western knowledge is a re-examination of the linguistic backgrounds of its thinking, and for that matter of all thinking." Whorf argues that the linguistic construction is

 
G. Mazzola et al., *Making Musical Time*, Computational Music Science,
https://doi.org/10.1007/978-3-030-85629-8_4

Fig. 4.1: Whorf argues that the linguistic construction is also strongly one of understanding and shaping reality.

also strongly one of understanding and shaping the reality, from mathematics to physics to the humanities. Especially when it comes to time, different languages have divergent temporal structures. He gives examples of the structure of verbs with Hopi language [165, Ch. 49]: "Verbs in the Hopi language are noteworthy for their very rich and expressive development of verbal aspects and voices.(...) Hopi also has three tenses: factual or present-past, future, and generalized or usitative. Hopi verbs belong to seven classes or conjugations having slightly different inflectional systems." Whorff states for another example, the language of the Chichewa, an unlettered Zulu tribe, that: "Chichewa has two past tenses, one for past events with present result or influence, one for past without present influence. A past as recorded in external situations is distinguished from a past recorded only in the psyche or memory; a new view of TIME opens before us.(...) It may be that these primitive folk are equipped with a language which, if they were to become philosophers or mathematicians, could make them our foremost thinkers upon TIME."

In their paper *Time, space, and events in language and cognition: A comparative view* [146], Chris Sinha and Peter Gädenfors state: "There is mounting evidence that not all languages and cultures employ space–time metaphoric mappings, and that not all space–time correspondences in language are metaphoric in nature. We also draw on our own and others' research to argue that the abstract conceptual domain which we call "time as such" is not transculturally universal, but the product of systems for measuring time intervals, and hence a sociohistorical construction." When discussing the temporal relational schema, they first relate their discourse to McTaggert's A- and B-series (see Section 2.6), and they rename these series as D- and S-time with a stress on semiotic aspects: "D" stands for "deictic", which in semiotics means that a sign changes its meaning with its user. The A-series is exactly this deictic or "shifter" type of a sign. D-time is the standpoint of presence, which changes its meaning each time one refers to it. "Now" has a different, shifting meaning with each of its usage. One should recall that the Greek "kairos" is also such a deictic temporal category of presence. As opposed to D-time, S-time, which in the Greek philosophy would be called "chronos", is the sequence time, the sequence of event points which are invariant in their usage. The World War I was before World War II, this temporal fact is not a function of my presence.

They then discuss lexical realization of D-time constructs, mentioning that Chinese have no tense or other languages have D-time marked on the noun rather than on verbs. Lexical D-time happens to be complex in the Yéli Dnye language, spoken on Rossel Island off the coast of Papua New Guinea. However, this language also relates D-time instances to S-time instances. In general, S-time is more cross-linguistically variable in its lexical expression and conceptualization than D-time. Consequently, it is more difficult to establish whether its lexicalization is common to all languages. They: "conclude from this brief survey that, although both D-time and S-time schemas are almost certainly transculturally present, there is considerable variation in the specific ways in which these are organized and expressed in different languages."

From their discussion of the temporal relational schema, they conclude that: "events may be more cognitively fundamental than the categories of space, time, and object.(...) We briefly present a model that is cognitively grounded, based upon the theory of conceptual spaces. An event can be described as built up from an *agent*, an *action*, a *patient*, and a *result*."

They then relate events to time: "It is important to note that this model of events does not explicitly represent the time dimension.

However, since actions and events are dynamic entities—they unfold over time—temporality is implicit in the model.(...) The semantics of verbs can be represented on the basis of this model without presupposing an explicit metric time dimension."

They then discuss the D-time (event time) intervals as opposed to S-time (elapsed time) intervals and refer the Greek duality of kairos vs. chronos. They conclude: "Event-based time intervals are intervals whose boundaries are constituted by the event itself. In this sense, there is no cognitive differentiation between the time interval and the duration of the event or activity which defines it, and from which, in general, the lexicalization of the time interval derives. For metrical time (S-time), calendars and clocks are cultural inventions that represent it. Not all cultures have calendar or clock time or employ metric time intervals. For example, in the Amondawa language and culture of Amazonia, there are only four numbers and lack calendric terms." The authors conclude: "The abstract "Newtonian fourth dimension" that we label "time as such" is not a universal conceptual domain, but a cultural model that has shaped our cognitive ecology.(...) Our argument has been that event representation is the fundamental basis of the human understanding of, and communication about, time."

In a concrete study [11], Panos Athanasopoulos, a linguist from Lancaster University, and Professor Emanuel Bylund, a linguist from Stellenbosch University and Stockholm University, asked Spanish-Swedish bilinguals to estimate how much time had passed while watching either a line growing across a screen or a container being filled.

At the same time, participants were prompted with either the word 'duración' (the Spanish word for duration) or 'stid' (the Swedish word for duration). The results were clear-cut.

When watching containers filling up and prompted by the Spanish prompt word, bilinguals based their time estimates of how full the containers were, perceiving time as volume. They were unaffected by the lines growing on screens. Conversely, when given the Swedish prompt word, bilinguals suddenly switched their behavior, with their time estimates becoming influenced by the distance the lines had travelled, but not by how much the containers had filled.

The researchers concluded: "The fact that bilinguals go between these different ways of estimating time effortlessly and unconsciously fits in with a growing body of evidence demonstrating the ease with which language can creep into our most basic senses, including our emotions, our visual perception, and now it turns out, our sense of time."

From this research it follows that language is strongly related to temporal concepts and the associated understanding of reality. But we do not have to follow Whorf in his thesis that language is causing the way of thinking. Nevertheless, one should become aware about the strong relationship between language and reality control. This relationship is also stressing the aspect of embodiment in temporal conceptualization, the index that events and in particular gestures are core in the creation of temporal culture.

### 4.1.1 Time in Literature

Time in literature, for example in Marcel Proust's *A la recherche du temps perdu*, and in narrativity theory, are important classical topics. We shall not discuss them here because we believe that they don't contribute essentially different perspectives to the phenomenon of time in music. But let us at least sketch the three time categories in narrativity theory: Gérard Genette's [43] distinction between 'story,' 'discourse,' and 'narrating'.

- *Story Time* It is the time, where the story takes place, say the historical time in a historical text. This type is always strictly chronological, in absolute time categories (years, months, days) or in an imaginary world of a fairy tale.
- *Discourse Time* It is the time it takes to read through the given story. It's the reader's time. It is not strictly chronological since the reading of a text might require rereading of parts, especially if the text is self-referential or just too difficult to be read and understood in a single run.
- *Narrating Time* It is—in a simplified understanding—the temporal relationship between story time and presence of the discourse time. When one reads about a story time, one also defines the temporal distance to one's own time, the time while reading the story, often using the past-tense.

**Impact.** In languages, time is conceived as a relationship among different types of events. The Newtonian abstract time line is often absent. Time is only implicitly addressed when events are formulated in their presence for human existence. This being, the impact of linguistic time concepts is dominated by what happens in the context of relevant events. Implicitly, this points to gestural categories, from which the temporal category results as a byproduct, and not as an a priori abstraction.

– $\Theta$ –

## 4.2 Time in Upbringing and Social Grouping

**Summary.** For the experience and control of time, music plays a key role. We discuss the development of musical abilities in upbringing of humans.

– $\Sigma$ –

Let us get off the ground by a summary of well known facts. In child development, music plays an important role, (mostly) mothers sing to calm and soothe children, and later infants recognize the melody of a song long before they understand the words. Toddlers love to dance and move to music. Children reproduce rhythms by clapping or tapping objects, see Figure 4.2. This gestural approach comes before any abstract acoustical participation. The key to toddler music is the embodied repetition, and this one in turn encourages language and memorization. This implies that the rhythmical embodiment is a root component of musical development, and this means that the embodied gestural construction of musical time is a fundamental component of musical development in humans.

Fig. 4.2: Children learn musical time via embodied gestural experimenting.

Moreover, a 2016 study of a research group around Assal Habibi at the University of Southern California's Brain and Creativity Institute [47] proved that musical experiences in childhood can accelerate brain development, particularly in the areas of language acquisition and reading skills. They conclude:

> In summary, we have shown accelerated development of auditory cortical potentials and greater ability to detect pitch changes in children who underwent two years of musical training compared to age-matched comparison groups. Taken together *these findings provide evidence that childhood music training has a measurable impact in the development of auditory processes.* Although the findings described here are restricted to auditory skills and to their neural correlates, such enhanced maturation may favor faster and more efficient development of language skills as well, given that some of the neural substrates to these different processes are shared. *Our findings demonstrate that music education has an important role to play in childhood development and add to the converging evidence that music training is capable of shaping skills that are ingredients of success in social and academic development.* It is of particular importance that we show these effects in children from disadvantaged backgrounds.

The role of music transcends the human maturation from the earliest years of life into a social factor of primordial signification. There have been many investigations about this fact, and aready Charles Darwin noted in 1859 that social instincts, including song, are the prelude for much of what governs our social evolution [32]. We refer to an excellent overview by Jay Schulkin and Greta B. Raglan [141]. They relate to our previous upbringing topic:

> Exposure to music in young children is known to promote prosocial behavior in children. Studies suggest joint singing or drumming, for instance, when controlling for diverse intellectual and personality factors, promotes prosocial behaviors. (...) Group size and social contact is linked to neocortical expansion in hominids, as is longevity. The pressure of coming into contact with others, creating alliances, and tracking them no doubt required more cortical mass. (...) A broad based set of findings in non-primates has also linked social complexity to larger brain size. Technology, including musical objects, is an extension of ourselves that expands what we explore, facilitating plasticity of expression and long term social bonds.

The conclude their study with a reference to Ian Cross [31]: "Ian Cross has noted that facilitating the transmission of information across shared social intentional space is the pervasive social milieu; evolutionary factors are critical in understanding musical sensibility, specifying diverse social contexts in relationships. We use music because it expands our communicative social contact with one another. We also enjoy music even without obvious instrumental features. Music, like other features about us, became a worthy end for its own sake."

It is remarkable that these authors stress the embodiment in music, not only its acoustical reality:

> Music is action oriented, whether literally in the movement or the virtuosity of a Liszt, or in the controlled building up to a crescendo and release as in "The Lark Ascending" by the 20th century composer Vaughan Williams. Action permeates music and dopamine[1] underlies the action of thought and the diverse cognitive systems that orchestrate the embodied expression of music.

## 4.3 The Musicality of Everyday Life

**Summary.** Music is an essential component of everyday life in most cultures (even where it is forbidden for religious reasons, its meaning is recognized in a negative understanding).

– $\Sigma$ –

Generally speaking, all cultures share a musical reality. The most frequent traditional presence of music is visible in rituals of all sorts: Singing to infants, initiations, marriages, funerals, religious rituals, solemn social events of official character, in sports, in military contexts of peace and war.

But in modern times with refined and globalized music technologies, music has also 'advanced' to a more or less agreeable 'noise' in everyday's life. We recognize the musical background of commercials, its omnipresence as muzac in malls, restaurants, and restrooms. And with the smartphones, everybody is carrying around one's own musical cloud to manage loneliness and the day's functional hotspots (waking up with music, having a sweet musical background in a private dinner, traveling by car, train or airplane).

Our everyday's life has become enhanced by an omnipresent musical dimension. Listening to music in these contexts has no longer the function humans were used to before the introduction of musical technology. Nevertheless, the ritualistic function is still present, although altered and trivialized. The musical roots of everyday's rituals have become a generator of one's existential color.

It would be a "terrible simplification" to reduce the function of music in cultures to the biological aspect of sexual attraction. The overarching role of music in cultures—beyond its sophisticated role in artistic creativity—is its driving force as a ritualistic motor, a drive that enriches one's rhythm of life. What is the nature of this generic role? Evidently, it cannot be exhibited by

[1] Dopamine is a central organizer of drives and rewards and is tied to music sensibilities imagined, acted, and expected. The regulation of dopamine is, for behavior, a fundamental event. It is an ancient molecule dating back millions of years in evolutionary history and plays a critical role in the motor control of the nervous systems of all vertebrates.

any specific style, be it harmonic, melodic, or rhythmical, the differences in different cultural contexts are too significant.

We argue that the common denominator could be the shared characteristic that any music generates an 'existential counterpoint' to life in profession and survival. This antithesis shares the construction of a temporal reality that one imposes upon the extramusical reality. Perhaps the best index of this function is music in movies. Take away the music from a movie, and it boils down to an insignificant visual patchwork. The commonplace that movie music is ideal, if one is not aware of it, this is a wrong statement: Music is the most basic component of a movie. In this understanding, life without music would degenerate into to a movie without sound. In other words, *the construction of a 'contrapuntal reality', and more specifically: a 'contrapuntal time', seems to be the core of musicality in everyday's life.*

**Impact.** The eminent role of music in upbringing and social life is proved from results in neuroscience, cognitive science, and psychology. The creation of a musical reality can be focused on the construction of a 'contrapuntal reality' as an alternative and guiding force against biological and physical contexts. This construction gets off he ground by a complex architecture of gestural events that embody time as an organization of social character. The acoustical expressivity of music plays a secondary role, it is a corollary of the event-based genealogy. In the social genealogy, the organization of a sequence of events, typically realized in social rituals, defines time as a structure relating events following templates of a social nature.

– $\Theta$ –

## 4.4 Geographical Location

**Summary.** This section presents a selection of approaches to (musical) time in different cultural locations. We have chosen the locations of Europe, Asia (including Chinese, Japanese, and Indian perspectives), Africa, and Americas (including South American and North American perspectives).

It is important to understand that our choice is by no means representative for these geographic locations. This is evidently impossible since the variety of time cultures is huge and even in local contexts not a constant expressivity. For example, when we discuss the European perspective, the focus on the history of time culture as driven by mechanical clocks and the parallel introduction of mensural notation is only one of the European phenomena that include a manifold of oral folklores (the Hungarian tradition, for example), which we cannot discuss in this chapters.

It is also risky to impose a theoretical classification upon time cultures, as was proposed by Thomas J. Bruneau [53] under the title of *chronemics*, "the study of human tempo as it related to human communication". This theory distinguishes between *monochronic* and *polychronic* time, the dichotomy between simple and multiple tasking, and where different countries are classified according to these options (Switzerland, USA, are monochronic, whereas Africa and South America are polychronic). This type of classification is not what we want to follow, we prefer to give proper descriptions of time cultures without (as much as possible) theoretical biases.

Our choice was nevertheless not arbitrary. We tried to select aspects within these locations which are not found in other places with the same intensity and profile. For example, our discussion of an African time culture follows the famous Ghanaian musicologist Kofi Agawu, who confirms that "rhythm is heart and soul of African music and its association with dance." Evidently, this aspect is quite distant from the also African Egyptian *maqam* music. Our choice is not a judgment for or against maqam music, Africa comprises a huge variety of time cultures, but Agawu's insight of rhythm being the heart and soul of African music remains a truth that one should investigate when sketching a geographic distribution of time cultures.

Summarizing, the reader should be aware that our choice of examples of time cultures in different geographic locations is not exclusive, and not a rating, but simply a choice of interesting and profiled phenomena. We encourage the readers to further investigate relations and developments of time concepts in different societies, which we cannot pursue in this book.

Our classification according to geographic locations is not meant to be a deep decision with regard to the culturally defined understanding of time (in music). But it nevertheless offers a spatial localization of time-relevant developments. The reader should therefore be aware of the cultural surface that a geographic specification might or might not entail. The geographic specification is simply a first perspective of time cultures without any ideological background. This is also confirmed by our choice of instances related to geographic positions: they are prominent examples, which show up when focusing on geographic loci.

– $\Sigma$ –

### 4.4.1 Europe

**Summary.** This section describes some roots of the European culture of time, which were essential when getting off the ground in the late Middle Ages.

– $\Sigma$ –

This perspective focuses on the simultaneous introduction of mechanical clocks and mensural notation of music, together with the consequences of this conjunction of historical events, which were characteristic for the development or European time in music.

#### 4.4.1.1 Time Was Introduced by Clocks in the Church

**Summary.** The European introduction of "Public" time was initiated by the installation of mechanical clocks (mainly on steeples). We follow and summarize the extensive discussion of this innovation in Rudolf Wendorff's excellent book *Zeit und Kultur* [164].

– $\Sigma$ –

Different types of clocks, using the sun's trajectory or water force were improved adding the verge escapement to weights around the end of the 13th century in Europe. This technological progress was soon spread everywhere, leading to an omnipresent visibility and (using bells) audibility of time machines on steeples and other important buildings in the increasingly important cities of the Late Middle Ages.

But not only this new technology was responsible for the increasing presence of time in Europe's society, it was also the increasing interest of Christians in the presence of one's individual life due to the weakening of the old eschatological expectations. Dante, in his *Divina Commedia*, was the first writer to mention mechanical clocks in 1320, in the book of paradise, song X:

cosi vid'io la gloriosa rota
    moversi e render voce a voce in tempra
    ed in dolcezza ch'esser non pò nota
se non colà dove gioir s'insempra.

In this spirit, the new time technology also improved the social, economical, and administrative organization of human interaction. This was also stressed by an increasingly refined subdivision of time intervals, from 6 daily portions to 12 and then to 24 hours, and also down the scale to minutes and seconds. The purely technological achievment however morphed to a more culturally determinant background structure. The flow of time was conceived as a divine setup. The mathmatician, philosopher, and bishop Nicolas Oresme in the 14h century compared the universe to a gigantic clockwork, built and maintained by God to generate a harmonic collaboration of its 'wheels'. It is remarkable that this idea of a divine time construction was later introduced to physics by Newton, who argued that space and time are God-given reference dimensions of all of physics, see also Section 3.1.

#### 4.4.1.2 Mensural Notation and Clocks

**Summary.** The coincidence of new mechanical clocks and the invention of musical mensural notation is discussed.

– $\Sigma$ –

It is remarkable that the introduction of the new clock technology at the end of the 13th century goes perfectly parallel with the invention of mensural notation in music. Its first accomplished presentation was first written by Franco of Cologne (*ars cantus mensurabilis*, around 1250). He distiniguishes four durations: *duplex longa, longa, brevis, semibrevis.* This notational progress was motivated by the necessity of having a precise collaboration among different, quite autonomous voices. So in music we also have an augmented desire of well-defined collaboration, which we already noticed above for the social layer.

The evolution of European music *sub specie aeternitais*, under the view of an eternal time line, which was also working following a divine clockwork, culminates in Johann Sebastian Bach's perfect machinery of the *Wohltemperierte Klavier*. Listening and watching Glenn Gould's performance of Bach reminds us of God's marionette hitting the piano's keys as driven by a higher string mechanism.

Although this European time technology is driven by gestural mechanical divices, such as Maelzel's metronome, the concept of time that it generates is a strong abstraction. Time is framed into a succession of time points (tic tac) that define start and end, without being concerned with the existential expressivity while the transition from start to end is happening. It is the opposite of the negative time, Ma, in Noh theater, see Section 4.4.2.2.

This spirit in European music has created an important tension between mechanical/metronomic time and the time, which stems from free gestural expressivity. This is manifest, for example, when we compare Glenn Gould's performance of Beethoven's Sonata op. 57 "Appassionata" to Vladimir Horowitz's rendition. Gould plays it as if it were a mechanical construction à la Bach, he misses Beethoven's underlying gestural approach, which is represented in Horowitz's stringent performance. It is also manifest in the same way when we compare Gould's interpretation to Maurizio Pollini's of Beethoven's Sonata op. 106, "Hammerklavier".

This tension and conflict remains omnipresent in European composition and performance to the present day. It has been traced in a quasi-ironic way in György Ligeti's composition *Poème Symphonique* for 100 metronomes, but also in the anti-metronomic compositions of Alexander Scriabin, for example his last poem op. 72 *Vers la Flamme*, which starts from large, slow durations and ends up with microscopic and fastest trill durations.

Ultimately, this conflict stems from the question of *who defines time in music*. Bach's mechanical/metronomic style expresses the divine clockwork time, whereas Beethoven's gestural expressivity insists on the human autonomy in the construction and perception of musical time.

**Impact.** The European introduction of musical time as a mechanical entity, enabled by the introduction of mechanical clockworks, was a strong contribution to he Newtonian divine time as an abstract reality where human activity would be embedded as an organization within an external ontology. This strong direction in musical time construction was however put into question by many important composers, such as Beethoven, Scriabin, or Ligeti. This conflict between an objective external and an individual internal approach to musical time is still virulent. But in the 20th century, especially with the introduction of improvisation in jazz and non-deterministic musical creativity, the 'counterpoint' of mechanical against individual, between divine and human time conceptualization has created in the Western musical culture a tension that developed an appropriation of musical time by humans without any essential reference to 'God's time machine'.

– $\Theta$ –

### 4.4.2 Asia

**Summary.** Asian culture deals with time as an expression of human existence, especially focusing on the moment of presence.

– $\Sigma$ –

#### 4.4.2.1 Understanding the Chinese Concept of Time From a Cultural Perspective

**Summary.** In the Chinese concept of musical time and spatial dimension, the notes are important, but more central to this genre are the spaces between the notes. The Chinese concept was later imported to Japan and developed as the "art of ma" with a temporal dimension. It can translate to multiple concepts, such as spacing, interval, gap, pause, rest, and timing. Western classical ideas of musical time emphasize the start and end of a sound as essential but do not highlight the inside of the event [123].

– $\Sigma$ –

The concept of time in Chinese musical tradition is different in the notation. In Western classical music, a person can hear one note, and there are a clear beginning and end. However, in Chinese music, there is a sound in between where a person can listen to the beginning, but not necessarily hear the ending (see *Chinese Gestural Music Notation* [102, Ch. 33.2]). In traditional Chinese music, there are two main notations, pitch, and technique. Pitch and techniques rely on the process of producing sound (how). Because of the lack of time instruction (or freedom) within Chinese tradition, the performers have to be masters of the time. In Western classical music, the composer already predetermined how each note (or the whole piece) lasted. The historical reason is that in early Western classical music, the medieval period, it was used in church choirs to have people sing together/unison. Having the exact time/rhythm is crucial to Western classical music (see *Western Notation and Gestures* [102, Ch. 33.1]). The precise time/rhythm is not the case in China, where there is only the solo master for entertainment and no need for having a time for others to follow in ancient China.

In addition to music, time fluidity plays an essential role in movement within the Chinese musical tradition. Similar to the notion of time fluidity in Chinese music, the movement and journey from a note to another are just as important as starting and finishing notes themselves. Both Chinese music and Western classical music notes may have a clear beginning and end. However, in Chinese music, the artistic process of getting to the end through articulation is different. Visual representation of this process is within Taiji ( 太极 ), an ancient Chinese philosophy about the natural world. The word Taiji itself refers to the "great primal beginning" of all that exists, and is often translated as the 'Supreme Ultimate,' and is an internal Chinese martial art movement. The critical concept of Taiji is fluidity, and the time concept in the Taiji movement is not counted by seconds and can provide gestures to the performers. On the outside, the movement looks slow; however, the amount of internal control, the muscles are continually working.

In Chinese philosophy, there is always a duality of subject and object (see Figure 4.3). In music, the performer is the music (composition), whereas, in Western classical music, the composer controls composition. Time is a gestural reality in Chinese music. The score is not just for reproduction (as a form of memory); it creates another reality by the specific qin ( 琴 ) – instrument master.

Fig. 4.3: Taiji Yin Yang.

Many Western classical musicologists think the symbol on a music score is the result and therefore is more substantial (overrated the music score) than the actual performance. Whereas in Chinese musical tradition, the symbol is not the result, and emphasis is placed on the action of the composer (fingers and instruments). When Chinese music score is translated into a Western classical music score, the action is lost. Chou Wen-Chung ( 周文忠 ) who was referenced in Chapter 5.3, his exploration of the integration of musical concepts and practices led to his ever-evolving theory of Variable Modes influenced by concepts found in the theories of Yin-Yang ( 阴阳 ) and the I-Ching ( 易经 ), Daoism, brush calligraphy, and Qin ( 琴 ), Chinese zither music, as well as early and modern European theories. He asserts that in Chinese musical tradition in reference to symbols [81, pp. 101-128], "the different aspects are not visible to the Western classical musician unless he is a scholar trained in his particular aspect, creates art that both are inventive, and resists cultural appropriation." Since Chou has experience in the Chinese musical tradition and Western classical tradition, he acknowledges the connections and distinctions.

The performance itself is more important than the score and can vary in time along with the tone depending on the musical tradition. The performers interpret the score and bring the symbols. The performance is constructed by these units, which are the "tones". Tone is important to the performance because the performance is created by tone. The tone can be different based on the performer's interpretation. The Chinese view a single tone as a meaningful expression through the use of timbre, rhythm, harmony, and most importantly, articulation. The playing of one-note acknowledges the existing harmony around it. Moreover, the performers' interpretation in the performance plays a crucial role rather than the score itself.

For example, Yan Pang's composition *Standing Alone*, which included Western classical and Chinese traditional influences, was originally composed for Alex Lubet on acoustic guitar. Lubet re-arranged on dulcimer to assimilate to the Chinese traditional music style. In the vibration of one string, rhythm and timbre resonate. After the re-arrangement, the time 'doubled' (time varies depending on the musical tradition) from 6 minutes to 12 minutes (refer to Section 15.2.1). These concepts are applied/used for Yan Pang's creation *Dancing Upstream* in Section 15.2.

#### 4.4.2.2 The Art of Ma Time in Noh Theatre

**Summary.** Noh theatre is about the art of ma, i.e., of negative space-time, the thing which is in between two events [68]. Noh is essentially moving in this negative existentiality.

– $\Sigma$ –

The Noh theater is essentially the "art of ma", of the negative time, the interval between two events. This approach to time reveals a culture of existential categories that are connecting events, and negating the isolation of singular events. Its negative characteristic is however not pure negation, it stresses their interconnection, in-between of events. The setup of a Noh narrativity stresses the transition, not the start and end points. This approach is a social statement about what happens when connecting events. When one participates in Noh theater, the events become secondary, and in fact, the Noh stories are secondary, the essence being the inter-space between the narrative's steps. The communication of this ma time is performed via the Noh's strong and explicit gestural forms. Ma is a gestural utterance, a form that dominates and overrides the narrative's contents, see Figure 4.4.

#### 4.4.2.3 Time Concept in Japanese Composer Joe Hisaishi's Music

**Summary.** We discuss the circular time concept in Joe Hisaishi's Music on two levels: the minimalist music and the ethnic approach to composition. The

Fig. 4.4: Ma in Noh theater is a gestural utterance, a form that dominates and overrides the narrative's contents.

music world is an intricate entity with many different languages to depict the condensing of the human spirit and different concepts of time. As individuals, we are always influenced by those music languages more or less.

– $\Sigma$ –

- **Intro to Circular Time A: Minimalism** Circular time are events in linear time. When one takes the musical gestures and repeats the same gesture again and again, similar to minimalist music, musical gestures are repeated. The repetition of the the action is important, instead of the musical content itself by creating a new dimension and focusing on something different. The repetition of the content is not only repetition, it is focusing on the action and creating another dimension. This creates *circular time*, not *linear time*. When you listen to the music, the repetition adds another level of attention and understanding. Listeners can forget about linear time and become immersed in an abstraction. The focus changed to a repeated gesture/action. When one listens to such music, they hear the identity and focus on the more abstract structure. It is a type of meditation, see also Figure 4.5 for an illustration of the meditative perspective being generated by a change of view onto the linear spiral.

  Prominent features of minimalist music include repetitive patterns or pulses, steady drones, consonant harmony and reiteration of musical phrases or smaller units. It may include features such as phase shifting, resulting in what is termed "phase music", i.e., process techniques that follow strict rules, usually described as "process music". This method is marked by a non-narrative, non-teleological, and non-representational approach, and calls attention to the activity of listening by focusing on the internal processes of the music. The repetition focuses on another reality. Composers, such as Arvo Pärt and Eric Satie, use this approach as well. This structured entity is not about the repetition, it is about the statement itself.

- **Intro to Circular Time B: Circulation Through/Between Tradition and Contemporary** Evolution of culture and history is a different way of looking at the circular time concept. This framework goes back many years and places traditional music next to contemporary music. This combination creates a new temporal identity. This revolution of music takes old structures and transfers them into the present. This is a higher level of minimalism, the repetition of cultural achievements. The present of the past is always there, eliminating a future directed, and evolutionary logic. Composers 'repeat' old structures to the present and thereby create a circular musical culture. The "present of the non-simultaneous," (or "co-present") is a classical trope in the theory of arts. This statement about the non-simultaneity is commonly accepted in painting; however, in musical composition, a similar paradigm was not known. Japanese composer, Joe Hisaishi ( 久石让 ) describes his musical approach as *minimal ethnic* [49].

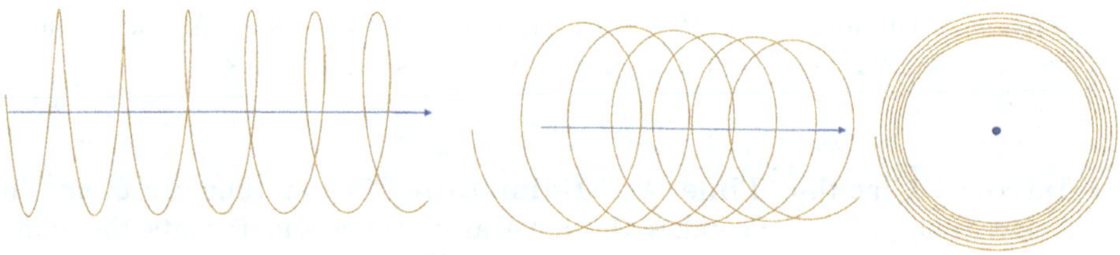

Fig. 4.5: Circular time is created from a spiral of gestures in linear time, but generating an added value when viewed vertically through linear time.

### Hisaishi's Musical Time Concept

- *Introduction to Hisaishi.* Some contemporary music developed much more decipherable languages that adopt many modern elements that might be closer to our lives. Hisaishi won countless music rewards and popularity around the world and changed the music history of Japan.
- *Minimalism.* Minimalism is another concept of time, an art form and a compositional practice that employs limited minimal musical materials. It does not necessarily follow a normal linear time span because of the repetition of musical motives/gestures. Modern interpreters seek the cultural significance of musical repetition "beyond the pleasure principle: repetition in music is thought to negate teleological desire, and thus repetitive music is allied with any psychic forces antithetical to *Eros*, to the goal-directed

patterns of tension and release that define the ego-creating life instinct." [41] In traditional music, repetition is a device for creating recognizability and reproduction to represent the ego. Within experimental musical circles, repetition is prized precisely for its ability to dissolve traditional formal dialectics, unleashing strange and unpredictable surges of intensity [41].

The music, *Summer* [152] in the film *Kikujiro* ( 菊次郎の夏 ), is an example of how Hisaishi adopts this concept of minimalism to his music. By repeating the same motif as the baseline, he structures his music to compose a strong backbone. His melody can be written simply and straightforwardly, which never loses its central idea while the notes in the melody are jumping up and down like a wild horse, which—impeccably portrayed—is the vivid summer. In the developing the composition and the accompaniment, Hisaishi alters the melody into different pieces while maintaining its structure. The right hand, which is the accompaniment, repeats the same pattern, altering and outpouring of fatality through music.

Another example of Hisaishi's minimalist idea, which is more explicit, is *Bolero* [15]. This piece's structure is much more evident for its repeating motives. The repeating motives are in the entire structure, not just in the accompaniment. You can hear the same thematic element repeating all the time, but the music is still moving forward instead of staying the same as when the composition started. Hisaishi uses this idea to create high intensity and tension. It makes the listener feel as if someone were anxiously running to an urgent matter. In this context, repeated phrases represent anxiety. This is how Hisaishi portrayed anxiety within his music using the idea of minimalism.

- *Circulation Through/Between Japanese and Western Influences.* Japan was Westernized after the Meiji Restoration ( 明治维新 ). Japan wanted to preserve its musical culture and saw the rise of hybrid Japanese-Western music as a possible avenue for preservation. This trend galvanized Hisaishi to write music into a combination of new American popular music concepts and Japanese traditional music. He deploys concepts from Japanese tradition and new age philosophies to address his different audiences. Understanding this complexity is important to understand Hisaishi's concept of time. This is evident in his work *Spirited Away*, which is widely popular and was a big moment of Hisaishi's glory.

  In *Spirited Away* the secret ingredients was the combination of new Western music concepts and traditional Japanese ethnic music. For instance, in *One Summer's Day* [122], listeners can feel how Hisaishi created a big space inside the music, an impact which dominates new age music. In the beginning, he uses a series of extended chords. After each chord, there is an ample space, adding mystery into the atmosphere. The existence of the extended harmony heats its tension that contributes to its mystery. When the melody pops up, the accompaniment remains in its initial state. Once a single chord sounded, it flew away with nothing but its echo. His music is about taking different times and cultural moments and putting them together. Hisaishi uses extension chords[2] associated with jazz, but it does not have a distinct jazz character in his music. Though inspired by jazz music, Hisaishi only extracts what he needs. He combines extended chords with the classical waltz. However, if this piece only engaged in contemplating new age music, it would not have received an Academy Award in 2003. Japanese traditional music elements fused with American pop music components triggered a glamourous wonder of animation music. During the Tang Dynasty ( 唐朝 ), 618-907, the Chinese music culture had spread to Japan. They began to use these Chinese scales. The problem during that time was that only a few Japanese individuals were enthusiastic about maintaining inherited consistency. The musical notation began to diverge as it spread throughout Japan. Then they developed a unique national music system of their own.

- *Hisaishi's Ethnic Markers.* The application of Japanese ethnic tradition is the core of the success of Hisaishi. Earlier, he composed music in traditional style and shocked people for decades. In 1988, the film *My Neighbor Totoro* jumped onto the front line of the viewers. Like always, this animation received praise for its impeccable quality. *The Path of Wind* was one of Hisaishi's masterpieces. This piece, one could say, is a modernized reflection of traditional Japanese ethnic music. The beginning is a series of 4th intervals which depicts the windy divine and mysterious forest lives. The beginning already established the necessary ambiance of the piece. When the melody appears, it is evident that the first part of the theme is in Japanese scales. The combination of Minyo ( 民 ) scale and Ryukyu ( 琉球 ) scale, see Figures 4.6, 4.7, is also a typical feature of Japanese music [145]. Chinese music refers to this combination as the Yu hexatonic ( 羽调六声 ) scale. Hisaishi endows the music with Japanese tonalities and with traditional instruments. His success is testament of the universal value of ethnic music.

Fig. 4.6: Minyo scale.

[2] An extended chord is an advanced composition technique. In short, the composer uses a chord more complex than the 7th chords (e.g.,11th 13th chord, even 15th). In the concept of jazz music, it is a chord adding extra notes for more color of the sound.

Fig. 4.7: Ryukyu scale.

"Japanese tradition" is also a dynamic and complex construction. "Japanese tradition" and "traditional" will be considered in the context of Hisaishi's music and popular culture. Still, it should be clear at the outset that tradition often becomes part of the personal and collective narrative in constructing a national identity. Culture, and therefore music, can play a primary role in creating the relationship between individual and nation and become the backbone of Hisaishi's music. He almost always includes ethnic markers in his scores. Not only in the path of the *Wind* and *One Summer's Day*, but also in many other animation pieces such as *Princess Mononoke*. Like he always did, he combined traditional features and Western music concepts.

**Impact.** *Cultural Significance of Circular Time.* What we need to understand is not Hisaishi's technique, but his ideology, which is always creative, bold, and patient. Traditional music might lie in the past, but it is accumulated in a long-term period of a nation, which has superior value in its deep core. We might have many new music elements today, still, they were the sediment of our predecessors' attempts to combine countless musical elements from the past to forge fresh sounds. A lot of common reasons for success are that they are able to combine things to present. Having longer time spans to work with without ignoring the ability to reference back in time. Circular time concept is fluid, the period of reference time can be variable. It is important for the composers to have this fluidity so they can create their unique voices through the gestural reference to earlier events and expression.

– $\Theta$ –

### 4.4.3 Importance of Indian Classical Music

**Summary.** Indian classical music has one of the world's oldest historical traditions and dates back to nearly two thousand years. It is unique in that it is characterized by a highly spiritual nature, as opposed to Western music cultures, whose mainly christian spirituality is not a persistent dominant factor. Indian music has developed through very complex interactions between peoples of different races and cultures over centuries. In a musical tradition in which improvisation predominates, and written notation, when used, is skeletal, the music of past generations is irrevocably lost. However, references to music in ancient texts, aesthetic formulations, and depictions or written discussions of musical instruments can offer clues. In rare instances an ancient musical style may be preserved in an unbroken oral tradition.

Time played a crucial role in this development as it is the synthesis of art, literature, science, mythology, religion, and philosophy which has improvised in different eras. It is impossible to divorce Indian music from the whole structure of Indian culture and philosophy with which it is interwoven in a number of ways from the earliest time of which we have records.

Our discussion of time in Indian music is provided by our Indian co-author Sangeeta Dey, her references are driven by her culturally genuine provenance, and not by scholarly standardized links. There are good books dealing with Indian time in music, e.g. Martin Clayton's *Time in Indian Music* [27], to which we may also refer. Clayton focuses on North Indian music. He stresses that the "development of *rag* is a sequence of gestures.", however without defining gestures in a non-intuitive way. A *rag* is not a score, it is strongly identified in its performance in time. Clayton's discussion of time does however not focus on the development of a *definition* of time, be it a human construction or a given ontological dimension. He rather focuses on the linear versus cyclic characters of meter and rhythm, where he strongly refers to Jackendoff's and Lerdahl's metrical theory (which we discuss in our Section 6.2.2) and somewhat to Jonathan Kramer's work (which we discuss in Chapter 9).

– $\Sigma$ –

#### 4.4.3.1 Chronology

- **Indus Valley Civilization(2500 BC-1500 BC)**
- **Vedic Music (1500 BC-500 BC)** Vedic religion was based on performing sacrifices in order to propitiate the gods. Music formed an important part of the rituals, which structured the sacrifice. In fact, singing, instrumental music, and dance were described as divine in Vedic literature; it was believed that they propitiated deities.
- **Music in Literature (500 BC-200 BC)**
  The first Indian epic, *Ramayana*, was composed by the sage *Valmiki*. It was written in *shloka* form. The epic tells us that musical instruments were collectively mentioned as *atodya*. Four major types of instruments were identified. Here are the basic musical structures:
  1. seven notes (*saptaswara*)
  2. three basic locations for tone-production (*sthanas*)
  3. four fundamental ways of empowering tonal arrangements (*varnas*)
  4. two basic intonation modes (*kakus*)
  5. six embellishments (*alankaras*)
  6. six aspects (*angas*).
- **Music and Natyashastra (200 BC-300 AD)** *Harivamsha* is a volume of 16,374 *shlokas* appended to the great epic *Mahabharata* between 200 BC

and 500 AD to complete the epic. *Harivamsha* is important because it describes two forms that may have inspired many composite genres in Indian cultural expression—the *Chhalikya*, a genre of songs in the ancient *Gandharva* mode of music making, and the *Hallisaka* dance. With its historical and deep-rooted religious tradition, Indian mythology holds music to be of divine origin. *Narada* was the first sage to whom the laws of music were revealed; *Tumburu* was the first singer; *Saraswati* was the goddess of music and learning; and *Bharata* was the first to draw up rules for theatre, of which music was a major and integral part. *Natyashastra*, or the Science of Theatre, a treatise on dramaturgy, is said to have been authored by *Bharata* sometime between 200 BC and 200 AD.

- **Moving towards Raga (300 AD-600 AD)**
  Hindustani music stressed improvisation which completely changed its nature. But the approach and concepts of *Dattilam* made the transition from *sama*-music to the contemporary *raga*-music significant and smooth.
- **Ragas,Talas,Tala-Music (600 AD-1200 AD)** *Brihaddeshi* (The Great Treatise on the Regional), by *Matanga*, was the first work to describe music in the period after *Bharata*, before the advent of Islam began to influence music. *Matanga* probably hailed from south India. *Brihaddeshi* is the first major and available text to describe the *raga*, which has been the central concept in Indian art music for centuries. It also introduced the *sargam*, or notation in the names of notes. In *Matanga*'s discussion of musical scales and micro-tonal intervals he clarifies what *Bharata* had said in the *Natyashastra*.
  One of *Matanga*'s major contributions is his scholarly focus on the regional element in music. '*Deshi*' has to be understood in contrast to '*Margi*' music, which is sacred and pan-Indian in its scope. According to *Matanga*, "*Deshi* is that which is sung voluntarily and with delight and pleasure by women, children, cowherds, and kings in their respective regions." *Deshi* music captured the flavor of a range of human emotions from different regions. Through notes it was formalized into ascending and descending scales.
  The present system of Indian music stands on two important pillars: *raga* and *tala*. *Raga* is the melodic form while *tala* is the rhythm underlying music. Together, *raga* and *tala* distinguish Indian music from many other musical systems of the world. The rhythm of music is explored through beats in time. Melody evolved as the *raga* through several processes; the *tala* resulted from a similar evolution in rhythm.
  Thus *raga*, which means color or passion, became a framework to create music based on a given set of notes (usually five to seven) and characteristic rhythmic patterns. The basic constituents of a *raga* can be written down in the form of a scale (in some cases differing in ascent and descent). By using only these notes, by emphasizing certain degrees of the scale, and by going from note to note in ways characteristic to the *raga*, the performer

sets out to create a mood or atmosphere (rasa) that is unique to the *raga* in question.

The idea of the *tala* is embedded in the concept of time. In Hindustani music it is the artist who bestows quality on time. A musician marks the beginning of his *tala* whenever he wants. He also creates his divisions in time. He thus creates the first beat. The flow of time is now released, channelled and directed. The artist then creates a beat to mark the first division or segment. With this first division in time the flow becomes comprehensible. The artist subsequently puts in successive and equidistant strokes. He thus makes available to us the *matra*, a measure to compute musical time. The duration between two *matras* is known as the tempo. The release of the time flow and the determination of the measure to compute it are the primary requirements to make a *tala*.

Cyclical and repetitive time-patterns composed of groups of long and short duration time divisions are *talas*, as we know them today. In every *tala* in Hindustani art music clapping (tali), tapping of fingers and waving of the palm (*khali* or *kal*) are analogous. These weave a pattern of sound and silence. Ancient treatises enumerate 108 *talas*. However, contemporary performances are normally restricted to about 15 *talas*.

*Talas* gain life and body when instruments play their role. Instrumental sounds, when expressed onomatopoetically, formulate sound syllables. These sound syllables, when fitted suitably to the *tala*-divisions, create *thekas*, the *tala*-expression that is actually played and heard in Hindustani music.

Thus the *talas* function as accompanying entities in Hindustani music and dance. They also serve as the basis for solo renditions in rhythm music.

Hindustani art music began to evolve after pre-medieval Indian music passed through certain stages of transformation and development till the beginning of the 11th century. Many Indian and non-Indian cultures took an active part in this transformation.

- **Music in Sultanate period (1200 AD-1700 AD)**

  The '*Ashtachap*' stream of music was established (1607-8 AD). It was named after the eight musical acharyas or preceptors who composed the music of the cult. The legendary *Tansen* too came under its influence. '*Haveli sangeet*' was the temple music practized by the '*Pushti Margi Sampradaya*'. *Nathadwara* in Rajasthan was the main seat of this *Vaishnava* devotional cult. The cult has created a rich historical tradition of temple-based music described as '*Haveli sangeet*'. '*Haveli*' is a temple visualized as a palace that the deity chooses to live in.

  The musical history of the post-*Ashtachap* period of *Pushti-sangeet* coexists with many developments in Hindustani Art Music The advent of the *Dhrupad*, *Khayal* and *Tappa*, the dissociation of dance from music, and the shift from the *pakhawaj* to the tabla, all happened during this period.

  India in the sixteenth century was politically and geographically fragmented. There were also multiple cultural forces at work. More than nine

rulers vied with each other to promote their own respective court cultures. Commoners were allowed freedom in matters like religion. In various courts a sophisticated court culture evolved and crystallized. This enabled the emergence of a chunk of art and classical music distinct from devotional or folk music. This court music exhibited a great deal of Muslim influence.
From the beginning of the nineteenth century many Indian scholars began to publish material on Hindustani music in English as well as in regional languages. This was a welcome addition to the works of the early British Indologists.
The term *gharana* is derived from the Hindi word '*ghar*'. This in turn can be traced to the Sanskrit word '*griha*', which means 'family' or 'house'. The *gharana* concept gained currency only in the nineteenth century when the royal patronage enjoyed by performers weakened. Performers were then compelled to move to urban centers. To retain their respective identities, they fell back on the names of the regions they hailed from. Therefore, even today, the names of many *gharanas* refer to places. Some of the *gharanas* well known for singing *khayals* are: *Agra, Gwalior, Patiala, Kirana, Indore, Mewat, Sahaswan, Bhendibazar* and *Jaipur*.
A *gharana* also indicates a comprehensive musicological ideology. This ideology sometimes changes substantially from one *gharana* to another. It directly affects the thinking, teaching, performance and appreciation of music.

#### 4.4.3.2 Essence of Spirituality

"Vibration was the first original aspect of Brahma, the creator. Every impulse, every action on any plane of existence has its origin in the one source." Inayat Khan, Sufi message.

The entire creation of God is based on the *Nada Brahma*. Mystics say that this creative sound is God Himself in dynamic action. God is the center of all existence. He is one, at rest, motionless, undifferentiated, silent and pure. His act of creation is really an act of projection or emanation. He makes waves within himself and the primary wave or motion from which all other motion is derived, and by which it is actively sustained, in his sound. It is also called his word, his cry, his voice, his call, his name, and by many other names. The *Vedanta* speaks of *Nada Brahman*, Sound-God, meaning that the word or sound or vibration was the creative aspect of God, which shows that the mystic does not differ much from the scientist who says that movement is the basis of the whole creation.

Some scientists and philosophers have come close to the descriptions of mystics when they have suggested that the fundamental reality of the universe is a sound or a motion that rings and flows through all things. Mystics have said that this sound can be known as reality through direct, personal mystic experience. In the Vedas, among the oldest of the world's writings, this creative sound is called the '*Nada*', which means sound, while other Indian mystics called it the '*Shabda*'.

The spiritual essence of *raaga* depends up on the rendition of particular *raag* by an artist, the lyrics, mood, concentration, and presentation of particular composition by an artist. Only then the musical waves touch the soul of an artist and get drenched into that particular *raag*'s essence (*rasa* and *bhav*). Then only the essence of spirituality of a *raaga* can be experienced by an artist and at the same time also a devotee. This shows that the *Aahad nada* (external music) helps the devotee, an artist, to reach the heights of *Anahad* sounds (Internal music) through concentration and meditation to some extent. This concentration and meditation can be attained in different kinds through *bhajana, nama smarana*, listening to devotional lyrics, chantings etc. These kind of attempts may help a devotee to reach some extent of spiritual heights. The *Aahad nada* that is the *raagas* which we sing and the classical music, the sounds and the rhythms that we hear are all *Ahat nadas.* But if an artist or a musician sings any *raaga* or any *bhajan* with pure devotion, *bhava ras*, then the listener or the devotee attains some concentration which also some times makes the devotee overwhelmed with tears. That is the power of music. This *Aahat nada* helps the devotee to concentrate and meditate to reach and experience the original eternal sounds that is the *Anahad* sounds which is an ultimate joy beyond human understanding.

The *raaga* is a melodic scheme. It is a nucleus based on certain traditionally accepted rules which in actual performance are improvised upon, expanded and embellished, thus, drawing out the possibilities inherent in the melodic embryo.

For Indian musicians groomed in the tradition, *raaga* is not grammar but a deity. The *raaga* is to be invoked and not simply presented with skill. The listeners, on the other hand, are expected not to treat the *raaga* merely as a performance of music to be applauded at the end. The first step is to allow the *raaga* to reach and move you, and the second is to allow oneself to respond to it. This is so because every *raaga* is a channel for the meeting of two or more minds sharing a love for music. A *raaga* may attract initially due to its pattern but its ultimate appeal is due to its emotional content or the *rasa* (sentiment) which activates it.

Since the inter-relation of the notes of a *raaga* determines its *rasa*, the drone is of great importance. The drone is the meter through which the frequency relationships of notes are established. The ancients with their simple scales and the hard-type of instrument could not have perceived in full the *rasa bhava* of a *raaga.* The drone instrument *tambura* (*tanpura*) figures in the history of Indian music only at a later period. The term '*raaga*' in the musical sense came to be used when it was found that a group of notes having specific frequencies roused a particular feeling with the background of '*shruti*', the emotional effect of *raaga* was clearly perceived and this marks an important landmark in the history of Indian music.

- **Tala or time measures**
Musical time in India, more obviously than elsewhere, is a development from the prosody and meters of poetry. The insistent demands of language and the idiosyncrasies of highly characteristic verse haunt the music, like a "presence which is not to be put by." The time-relations of music "are affected both by the structure of the language and by the method of versification which ultimately derives from it," says one student of Indian music from the West. Until the nineteenth century, there was practically no prose in India and everything was learnt through the medium of verse chanted to regular rules. Both in Sanskrit and in the vernaculars all syllables are classified according to their time-lengths, the unit of time being a *matra*. Very short syllables of less than a *matra* also occur. Great stress has always been laid by Indian grammarians upon giving the 'exact value' to syllables in verse; and as there is no accent at all in Indian verse, the time-length is all important. This may account for the great development of time-measures in Indian music. The different time-measures for verse are most carefully laid down and have to be strictly adhered to. When grammar, philosophy, history and geography are learnt in verse, one gets the sense of duration and rhythm highly developed, and *it is this sense of duration that is the central thing in Indian time.* Any one who studies Indian prosody can easily see the great difficulty, to say the least, of obtaining a pleasurable result by combining Indian verse with Western tunes. One of the most difficult things for the foreigner to get away from in an Indian vernacular is the stressing of syllables. The division into words is not at all important in Indian verse, and so music does not take particular note of this. *In India words are more often set to music, rather than music to words.* It is easy to see then the importance of time-measure in Indian music. The Westerner often finds these time-measures far more difficult to master than the melodies, strange though those often are. The varieties of time-measure may be somewhat imperfectly realized by listening to the rhythmic beats of the drum in some distant village on a quiet moonlight night, when all other sounds are stilled and one can get the full benefit of this one sound. Sometimes one hears beats arranged in bars like this:
|2.1.1| 2. 1.1|2.1.1|2.1.1.|
|1. 2.|1.2|1.2|1.2|1.2|
| 2. 3| 2. 3| 2. 3|2.3|2.3|
|7.1. 2| 7. 1.2| 7.1.2|7.1.2|
Such an exercise will not only help one to appreciate the rhythmic soul of India and the intricacy of Indian time, but will also help to pass the hours when one is forced to lie awake. Though the nomenclature varies, as might be expected, the theory of *tala* (as time-measure is called) in the north and south is more uniform than that of *raga*. As usual, a fanciful origin must be found for *tala*. It is said that Bharata discovered the thirty-two kinds of *tala* in the song of the lark. Raja S. M. Tagore says that the word *tala* refers to the beating of time by the clapping of hands. Sometimes it is also

done by means of small handcymbals, which are called *tala* or *kaitala* or *kartell* (handcymbals). It may be that, as has been suggested, the main difficulty for westerners in realizing and enjoying the nice distinctions of Indian rhythm is that they have not acquired the habit of resolving mentally every unit into its constituent elements, so that they could sing them at a moment's notice.

Indian rhythm moves in *avartas* (bars) broken up into *vibhagas* (beats), each of which contains one or more talas. We can equally say of ours that it moves in sections broken up into bars, each of which contains one or more beats. In what does the difference between the two systems consist? It may be answered that theirs is derived from song, ours from the dance or the march: that both are based on the numbers two and three, but that they add and we multiply in order to form combinations of these. But the answer which goes deepest is that their music is in modes of time (as we saw that it was in modes of tune), and that ours changes that mode at will, principally by means of harmony. In order that rhythm, an articulation of the infinite variety of sounds, may be upon some regular plan, the plan must have some recognizable unit of measurement. India takes the short note and gives it, for a particular rhythm, a certain value as opposed to the long; Europe takes the stressed note and gives it in a particular rhythm a certain frequency, as against the unstressed, and graduates its force. We find the unity of the rhythm in the recurrent bar (which is always in double or triple time, just as our two melodic modes are either major or minor), and have to look elsewhere for the variety; they find variety in the *vibhaga*, whose constitution is extremely various, and must look elsewhere for the large spaces of time: they find unity in the *avarta*, and we find variety in the sections. Indian rhythms have their raison d'être in the contrast of long and short duration, and to identify these with much or little stress is to vulgarize the rhythms. Stress pulses and demands regularity; duration is complementary and revels in irregularity. In order to get the true sense of duration we have to get rid of stress. The value which Indian music attaches to time may be judged from a description of a certain musician as an excellent timist, and from the name of sextuple *Govinda Nair* given to a musician of *Travancore*, on account of his great skill in singing in sextuple accelerated time. One can hardly imagine such terms being used in the West. Musical time is based upon the *akshara* or syllable. Five main note lengths are recognized, made up of a different number of *Aksharas*. They are:

- Anudruta 1 akshara 1/4 matra
- Druta 2 aksharas 1/2 matra
- Laghu 4 aksharas 1 matra
- Guru 8 aksharas 2 matra
- Pluta 12 aksharas 3 matra
- Kakapada 16 aksharas 4 matra

One *avarta* or section contains from 2 to 4 bars or *vibhagas*, each of which is constituted by a number of *angas* (members), consisting of one or more of these time units. The *virama* or rest is used for lengthening the *druta* and *laghu* by any fraction. Each *avarta* must begin the time-measure correctly, and all the various time elaborations must be worked out in the *avarta*. The *SaiigJta-Ratnakara* gives 120 examples of different time-measures, formed by the combination of these time units, the bar varying in length from one to nineteen notes. Most of these are very unlike any of the *tala* employed today, and so there is nothing gained by discussing them. We shall therefore take up the time system as it is today in both Northern and Southern music. Here again there is a good deal of difference between the north and the south. Many of the times are the same, but the names and the method of classification are different. As with *raga*, so here also, the south has a very much more systematic classification than the north. According to Carnatic music, there are seven *talas*, each of which has five *jatis* or classes. The five *jatis* are classified according to the number of *aksharas* in the principal *anga*. These are said to correspond to the five castes, and their origin is traced to the five faces of *Isvara*. *Nataraja* (Siva) is supposed to have worked these out in his wonderful dance, while Brahma played the handcymbals and Vishnu the *mridahga*. This would certainly have been a band worth going far to see and hear. The five *jatis* are named after the number of *aksharas* in the principal beat, viz. *trisra* for three, *chatusra* for four, *khanda* for five, *misra* for seven and *sahkirna* for nine. It is interesting to see that, with the exception of the second, all the other numbers are odd, the times being mostly combinations of two and an odd number. The same thing is found in Hindusthani tala. The other *arigas* of the *avarta* have either one or two *aksharas*. The following is a table of the *talas* as they are arranged in the Carnatic system. It will be noticed that they are not arranged in the order of the number in the principal beat, but in the usual Indian method of arranging them.

| **name** | **chatusra** | **trisra** | **misra** | **khanda** | **sankirna** |
|---|---|---|---|---|---|
| Eka tala | 4 | 3 | 7 | 5 | 9 |
| rupaktala | 2.4 | 2.3 | 2.7 | 2.5 | 2.9 |
| jhampatala | 4.1.2 | 3.1.2 | 7.1.2 | 5.1.2 | 9.1.2 |
| Triputa tala | 4.2.2 | 3.2.2 | 7.2.2 | 5.2.2 | 9.2.2 |
| Madhya tala | 4.2.4 | 3.2.3 | 7.2.7 | 5.2.5 | 9.2.9 |
| Dhruva tala | 4.2.4.4 | 3.2.3.3 | 7.2.7.7 | 5.2.5.5 | 9.2.9.9 |
| atatala | 4.4.2.2 | 3.3.2.2 | 7.7.2.2 | 5.5.2.2 | 9.9.2.2 |

The table shows that there is only one *anga* in each *vibhaga* for *ekatala*, in *rupaka* here are two, in *jhampa, triputa* and *madhya* there are three, in last two there are four. The *avarta* as we have seen is made up of a number of *vibhagas.* One of these take the principal beat and one of them has no beat at all. These beats are very important and the musicians have to keep them in mind, otherwise the time will go astray. In Hindusthani music the time-measures are arranged somewhat differently. We have first *eka tala* of the *chatusra* variety, and none of the other *jatis* are used. In *rupaka tala*, only the *chatusra* (2. 4) and *trisra* (2. 3) *jatis* are found. There is also another kind of *rupaka tala* which has three *arigas*, thus 3, 2. 2. *Jhampa tala* in the north runs 2. 3. 2. 3. —a kind of *rupaka.* There is also another kind of *jhampa* which goes 3. 3. 2. 2. In *triputa tala* we find the *trisra* and *chatusra jatis.* The former goes 3. 2. 2 and is called *tevra.* The latter is called *tltala, tJntal,* or *tritala*—three-beat—and also *kavali* in Bengal. There is also another kind of *tJtala* which goes 4. 4. 4. 4 with the *sam* on the third beat. *Madhya tala* is represented by its *chatusra jati,* which is called *sulaphakata tala* or *surphdkata,* meaning ‘zigzag’. Sometimes it runs 2. 4. 4 instead of 4. 2. 4. *Dhruva tala* is represented by its *chatusra jati* which is called *ata-chautala* and has two forms : 4. 2. 4. 4 and 2. 4. 4. 4, the sam being on the first *matra.* The word *ata-chautala* means “crooked four-beat time.” This time is used a great deal in *dhrupads. Ata tala* has three *jatis* in the north, viz. *chautala* 4. 4. 2. 2., *jhampa tala* 3. 3. 2. 2., *dhamar tala* 5. 5. 4. There are also a number of times which correspond to none of the regular southern times. These include the following : —. 2. 2. 4. 4 or 2. 2. 2. 3. 4. *Dhima tala,* also called in the north *adi tala,* 4. 4. 4. 4, with the *sam* on the first beat. *Dadra,* also called *pashto*—a syncopated *tala* especially used with the *dadra* class of song. It runs 3. 3 with the *sam* on the first note. *Jhumra.* 3. 4. 3. 4. It has the same on the first beat. This is a very popular time-measure.

Most of these irregular times were introduced by the Muhammadans. There is also an irregular southern time called *chapu tala.* It has two varieties, either *trisra* or *misra,* viz. 1. 2. or 1. 2. 2. 2. *Chapu tala* is used a great deal in folk songs. It is, as may easily be seen, a syncopated time. South Indian time experts reckon that there are altogether 108 different varieties of *tala* possible by means of various ingenious combinations. Most of them, it need hardly be said, are not in use, though occasionally some expert technician accomplishes a tour de force away from the beaten track of time-measures; as when Subramanayyar sang a piece in the *Simhanandana tala,* which is one of the most complicated. The bar signature of this *tala* runs as follows : 8. 8. 4. 8. 4. 8. 2. 2. 8. 8. 4. 8. 4. 8. 8. 4. 4 or 100 *aksharas.* This musician it may be surmised was a prodigy in South Indian time experts. As in Western music, so in India, it is possible to include in the *akshara* two, three, or even more shorter notes. These are called *kalai.* Their inclusion does not alter the time, but renders the singing or playing the piece more difficult. It means that the longer notes may be broken up into shorter ones, and so on, untill the ability of the performer is exhausted. A singer of *Travancore* was known as *Shatkala Govinda Marar,* because he could sing anything in sextuple time. In all this manipulation of time-measure, the main elements of the time must be retained throughout and the *raga* must be adhered to. The clapping of the hands is much used in India to indicate time. There are different signs used for the different beats, so as to make quite clear what kind of time is used. The first note of a beat is indicated by a clap. This may be followed either by the counting of the other *aksharas* with the fingers or by a wave. If the beat is a *laghu*, that is one of more than two *aksharas*, then the other units are shown by counting with the separate fingers on the palm, of the other hand. If the beat is a *druta* of two *aksharas* only, then the extra unit is shown by a wave of the hand.

In Indian music the graduated intensity of the sound is very little regarded, either in singing or playing or in drumming, because their whole scheme is not accentual but quantitative. It is true that the first of the bar is often louder than the rest, but not always; but this is not in order that it may, as with us, stand out against other accents; but because two quantitative schemes are apt to coincide there, and two sounds are louder than one. The time intervals are with them all important, and show great variety; it is seldom that more than a few bars, out of hundreds, are drummed in exactly the same way. And the drumming is practically continuous; it is only occasionally silenced for special contrast. The pitch again is all important, for it is invariably the keynote, and frequently the drum is the singer's only accompaniment. Lastly, a maximum of variety is got into the quality; and this not mainly by the variety of the instruments. For though there are scores of shapes for drums, tambourines, cymbals, triangles and so forth, they are not usually assembled together, because concerted music is the exception, not the rule. The variety is got out of the drum, or the pair of drums themselves. They are played with the full hand and the fingers, rarely with sticks; there are half a dozen strokes for the right hand and three or four for the left. Of these Lady Wilson's drummer said, "The beat with the left hand is like the seam of my coat, that must be there; the other notes with the right hand are like the embroidery I may put according to my own fancy over the seam." These 'notes' are differentiated not by pitch, but by quality. They are also articulated by great intricacy of time-interval. For neither of these two things has our music any real analogues; and the Bengalis do not overstate the case in their saying, "*Yantrapatir Mridanga*" (the drum is the lord of instruments). These two are the same in principle and are the drums used throughout India for the accompaniment of vocal music. The *mridanga* and tabla are both played in the same way, the only difference being that, in the case of the tabla, the two heads are on two small drums and not on the same drum. The right hand note of the drum is the keynote—the *Shadja*—and the left hand note a lower *Pa*. Exact tuning is very important, as the slightest difference will be evident and will spoil the melody, the drum being the principal accompaniment for the singer.

Drummers have a curious system of mnemonics, which tell them how the drum accompaniment should be beaten out. These are composed of syllables, each of which indicates one particular kind of beat and also the nature of the tala as a whole. The actual syllables used vary in different parts of India but the following are some of them:

Northern Tritdla 4.4.4.4.

+I Ta Dhin Dhin ta | Ta Dhin Dhin tal Ta Tin Tin ta |Ta Dhin Dhin ta|

Rupaka 3.2.2.

+|Dhin Dha Trik| Dhin Dhin| Dha Trik

Jhampa 2.3.2.3.

+I Dhin na| Dhin Dhin na| Tin na |Dhin dhin na

Chautdla 4.4.2.2.

+ indicates the *sam* and the *khali* beat. In these mnemonics, or *bols*, as they are called, the following are to be played by the right hand: *Dhin, Na, Ta, Trik, Tin, it, ki.* The following by the left hand: *Dha, Ta.* The following are played by both hands together:

The southern arrangement is somewhat different and runs thus: Aditala 4.2.2.

1 0 2 3

Ta ti Nam Tom | Ta—by the left hand with four fingers.

ti—by the right hand with four fingers on the middle of drum.

Nam—by the left with all fingers.

Tom—by both hands with all fingers at once.

+ 0 2

Tadimi Takitta Tam i.e. 2-2.4

Ta—by right hand first finger on the border of drum.

di—by the left middle finger.

mi—by the right middle finger on the middle of drum.

ki—as mi.

tta—as ta. Tam—by both hands simultaneously.

+ 0 2 3

Takitta tikitta Tonkitta Namkitta

Ta—by the left with all fingers and right with forefinger on the border,

ki—by right with middle finger,

tta—by right with forefinger on border,

ti —by left with four fingers.

Ton—by both hands with all fingers.

Nam—as Ton.

Rupaka 2.4.2.4

0 1 |0

Talangu Tom | Talangu Tay

Ta—by left hand.

Ian—by right with second finger on border,

gu—by middle finger.

Tom—by both hands with all fingers.

Tay—as Tom.

- **Ragas Have Specific Timings**
  The timings of ragas in Indian classical music is a very unique and distinguishing characteristics of both Hindustani (North Indian) and Carnatic (South Indian) schools of music. There are certain seasons and times of the day and night associated to each *raga*, and it is believed that the *ragas* are at the zenith of their harmony and splendor at those times. For example, *raga Bhairav* should ideally be performed at early morning, *Bhimpalashi* should be sung during late afternoons, and *Darbari Kanada* after midnight. If *Darbari Kanada* is rendered at forenoon, it is not going cast its engulfing magical charm as much as it would do after midnight
- **Bodily Elements (Humors) and *Ragas* in Indian Classical Music**
  The traditional Indian medical science, or *Ayurveda*, believes that our body has three *doshas* or evil elements—*vayu* (wind), pitta (bile) and *kof* (phlegm). These are quite similar to the four humors of Hippocratic theories of ancient Greece. According to Ayurveda, these elements work in a cyclic fashion through day and night in our body, each getting an upper hand at some point of time. Again, their effects vary from season to season; one happens to be more phlegmatic during the winter. Now, in Indian classical music, *ragas* have definite tones, which can in turn stimulate definite moods and sentiments. Thus, each *raga* arouses a certain kind of sentiment, working in connection with different humors or elements. We know music has the power to cure your depressions or insomnia; their connection with the bodily fluids or humors explain why. If you are melancholic, listening to a certain *raga* can make you feel sanguine by stimulating the concerned elements. And if listen to it in the prescribed time when the concerned humor is at its height (and the power of the *raga* is at its height too), the effect will be maximum.
- **Play Timings of *Raga* in Indian Classical Music**
  The *Thaats* of respective *ragas* are mentioned inside the bracket. Nevertheless, many experts may slightly differ on their views of the timings of *ragas* in Indian classical music. Thus, some may say that a certain *raga* should be instead at night instead of late evening, and so on.
  - After midnight (12 midnight - 2 pm): Adana and Darbari Kanada (Asavari), Malkauns (Bhairavi), Shahana (Kafi).
  - Late night to dawn (2 - 4 am): Paraj (Poorvi), Sohini (Marwa)
  - Dawn to early morning (4 - 6 am): Lalit (Marwa), Bhairav (Bhairav)
  - Early morning (6 - 8 am): Bhairav, Ramkali, Jogi (All three belonging to Thaat Bhairav)
  - Morning (8 - 10 am): Ahir Bhairav (Bhairav), Bilaskhani Todi and Komal Rishabh Asavari (Bhairavi), Todi (Todi)
  - Forenoon (10 am - 12 noon): Alahiya Bilawal (Bilawal), Bhairavi (Bhairavi), Deshkar (Bilawal), Jaunpuri (Asavari)
  - Afternoon (12 noon - 2 pm): Brindavani Sarang (Kafi), Gaudiya Sarang (Kalyan), Shuddha Sarang (Kafi)
  - Late afternoon (2 - 4 pm): Bhimpalasi (Kafi), Multani (Todi)

- Dusk (4 - 6 pm): Patdeep (Kafi), Purvi (also spelt Poorvi, Thaat Purvi), Puriya Dhaneshree (Purvi)
- Evening (6 - 8 pm): Hamreer, Shuddha Kalyan, Eman / Yaman / Kalyani (all belonging to Thaat Kalyan), Puriya (Marwa)
- Late evening (8 - 10 pm): Desh (Khamaj), Durga (Bilawal), Kedar (Kalyan), Jaijaiwanti (Khamaj), Pahadi (Bilawal), Shankara (Bilawal)
- Night (10 pm to 12 midnight): Bihag (Bilawal), Bageshri (Kafi), Chandrakauns (Kafi), Malhar (Kafi).

Some *ragas* like *Piloo, Dhani* and *Kafi* can be performed any time throughout the day. And as we have already written, different *ragas* stimulate different emotions. We are planning a hub on Indian Classical Music *Ragas* and Human Emotions.

- **Relation of Seasons with *Ragas* in Hindustani Classical Music**
  *Ragas* in Hindustani classical music relate to different seasons as well. Seasonal ragas can be performed any time (day or night) during that season. Summer in India is not as pleasant as in Romantic English Literature; its scorching heat is reflected through the *Raga Deepak*. *Megh* and *Miyan ki Malhar* are for the rainy season; legends say that it would rain whenever Akbar's court poet, Tansen, used to perform *Miyan ki Malhar*. *Malkauns* and *Puriya Dhaneshree* are the *ragas* for autumn and fall. The roughness of winter is conveyed best through *Bhairav*, another name for *Siva*, the god of destruction. And spring, we all know, is the season of vitality and romance. *Raga Hindola* offers the right melody to express that.
  The sounds that emanate in the different spiritual regions of the grand division in creation have similar earthly sounds through the musical instruments like Veena, flute, Been, Mridang etc. These descriptions are found in the Holy Books. Therefore *Ragas* and Spirituality are inseparable elements. Spirituality is the highest level, spirituality is Supreme Being and Supreme Being is *Rasa* (Brahmanand) without rasa nobody can imagine a *raag*. The tempo or speed of a musical composition in applied music must be in consonance with its *rasa*. Pieces in *raudra rasa* and the *vira rasa* must be *druta laya* or quick tempo. Pathetic pieces must be in *vilambita laya* or slow speed. Intricate time-measures are out of a place in applied music. Some *talas* are mentioned as suitable for particular *rasas* in ancient works.

**Impact.** The Indian time culture in music is loaded with a deep semiotic weight that connects time to categories of divinity. The *tala* system of rhythm stresses duration against accents of loudness. Time is not structured by other parameters (such as loudness), it is an immanent existential category. The complexity of metrical subdivisions enforces a strong focus on duration qua duration, time is an auto-referential dimension. It is however very different from the neighboring Chinese approach since its quantitative complexity steers away from the inner quality of Chinese temporality. But at the same time it is not the European extremism that stresses "on" and "off" without being 'interested' in the inner vibration. The realization of the Indian temporal musicianship is in fact not an abstract description of time intervals on the time line, it is a focus on the musician's complex performing gesture.

– $\Theta$ –

### 4.4.4 Africa

**Summary.** The African culture of time, as discussed by Kofi Agawu in [9, Ch. 4] is discussed.

– $\Sigma$ –

Fig. 4.8: Time concepts in African cultures are often developed and investigated untder the umbrella of rhythm.

Time concepts in African cultures are often developed and investigated unter the umbrella of rhythm, see Figure 4.8. Agawu confirms that "rhythm is heart and soul of African music and its association with dance." He however critizises that scholars are reducing this understanding to a "bodily knowledge". He proves in his discourse that African rhythm expresses a "rational complexity". He stresses the difference to European rhythms in the emphasis and idiomatic preference. He then exhibits three prominent features: time lines, polyrhythms, and lead drummer's narrative.

For our concern here (also check our discussion of the social genealogy of rhythm in Section 6.1.2), the cultural aspect of time is the focus. The first feature, time lines, is characterized by a multitude of musical times: reversed, discontinuous, sliding, circular, spiral. Often time-reckoning is related to social activities. As opposed to African languages, where time concepts rather sparse, musical time shows a sophisticated differentiation. Agawu concludes that musical time in Africa is not a microcosm of ordinary time, it is its own time realm.

This entails that the African culture of time in its specific and intense social reality generates a different time reality, a specifically musical one, and one that is differentiated in three modes of rhythmical expressivity: in speech mode, where the rhythm is not metrical, but driven by the semantic vector, in signal mode, where speech is stylized in time, and in dance mode, where the rhythm of the body is dominant. The social aspect of these modes is that many skills are combined in one philosophy, combining laymen's and professional's capabilities in an intelligent design of musicians' collaboration. This is explicit in the polyrhythmic structure of African musics, where many, eighteen or more, rhythmical lines are combined to a polyrhythmic maze.

Following Agawu's analysis, this complex collaborative structure of time lines, polyrhythms, and lead drummer's dimensions is a social construct of polyphonal, polytheistic, polycentric, or polygamous social organization. And moreover, this social construct is also strongly gestural, as realized by dancing, drumming, and singing movements.

A funny anecdote might illustrate the omnipresence of the dancing reality in Ghanaian society. When Guerino Mazzola visited the city of Kumasi in Ghana, they celebrated his wife Ida's birthday with a party of her friends in a good hotel. When Guerino had introduced the guests to this celebration at the dinner table, there was a short kind of embarrassing moment as nobody seemed to know how to make the next communicative step. In that moment, Ida stood up and started making dance movements without any further words. This spontaneous gesture was immediately understood and everybody started kind of dancing, while still remaining seated to the dinner table. But this dancing message immediately resolved the slight embarrassment, and we could start the dinner. What a strong expressivity of musical culture.

#### 4.4.4.1 The Sankofa Allegory

In Ghana, a famous allegory, the Sankofa bird (see Figure 4.9) embodies a symbol of circular time culture. The traditional saying about Sankofa states "Return to the source and fetch." Meaning "Go out, do something, and return with what you have achieved for the good of the whole community." [10] Especially in the arts it means that the source is the origin of art, of life, that they find deep down within themselves, and that they, as artists, can only bring forth as images and objects. Sankofa, looking back in time understands/recognizes the future (the egg). This is a circular time culture where only a full circle of temporal dimensions can yield a good equilibrium of our actions.

Fig. 4.9: The Ghanaian Sankofa bird representing circular time.

**Impact.** The dominating social aspect of these African time modes is that many skills are combined in one philosophy, combining laymen's and professional's capabilities in an intelligent design of musicians' collaboration. This is explicit in the polyrhythmic structure of African musics that are combined to a polyrhythmic maze. And the collaboration also includes the synthesis of past and future, as represented by the Sankofa allegory.

$-\Theta-$

### 4.4.5 Americas

**Summary.** This section deals with a sketch of US-American vs. South American time cultures that turn out to be radically diverse.

$-\Sigma-$

#### 4.4.5.1 North America

**Summary.** This section deals with a short view upon US-American perspectives of time cultures. We do not deal with jazz here, but with more industrial aspects.

$-\Sigma-$

Fig. 4.10: Harold Lloyd hanging on a clock in his movie *Safety First*.

The United States is one of the world's fastest countries, partly due to the fact that many Americans are always looking to the future, striving for the "American Dream". The commonplace "time is money" excels busyness, which identifies a hectic and frenetic life-style with success and status.

The fact that time is valued by its commercial consequences, its monetary equivalent, proves that time as such is inexistent, only its commercial 'shadow' is. As opposed to Switzerland, which has a strong time culture and technological tradition, the United States do not cultivate time as a proper value. When they say that "at the end of the day" (literally or metaphoric), something results, they also stress that the day as such is secondary, only its resulting outcome matters.

Perhaps the most emblematic image of the tyranny of running industriaized time is Harold Lloyd hanging on a clock in his movie *Safety First* (1923), see Figure 4.10.

#### 4.4.5.2 South America

**Summary.** Aymara: a language in which your past lies in front of you.

– $\Sigma$ –

The Aymara people, approximately 2,3 million in number, are an indigenous nation inhabiting the Andean highlands of western Bolivia, southeastern

Peru, and northern Chile. Over the last 50 years, the Aymara language has drawn a considerable amount of attention from researchers due not only to the peculiar way in which they describe time by means of space metaphors but also for the gestural, embodied components of these metaphoric relations. While it is known that that all languages so far examined take their vocabulary

Fig. 4.11: The Aymara language entails a sense of passage of time that mirrors our common sense.

of time primarily from that of space [120, p. 1], the Aymara language entails a sense of passage of time that mirrors our common sense. In Aymara, the past is located in front of the subject, while the future is placed in his or her back. This represents a reversing of the common metaphor for the perception of the order of future and past events and their spatial placement in the subject's bodily conceptualization of time. In all languages that have been analyzed so far, Aymara being the sole exception, the metaphor for time in relation to the ego—the experience of time centered on the body of the subject, not oriented towards series of events that are independent of his presence—maps future onto a metaphoric space that is located in front of the subject, and past behind the subject.

Let us take English as an example: it is very likely that, when asking an English speaker to describe events in a far future, like asking students who recently graduated from college when they are going to retire, that they would stretch the full extension of their arms in front of them to point to this unclear moment that might wait for decades to arrive. For a native speaker of Aymara, this spatial relation occurs in the opposite direction: they would put their hands as far as possible from their chests to describe their Inca ancestors, then somewhat closer to their bodies when talking about their great-grandparents.

Aymara vocabulary concurs with this unique conceptualization of the organization on this back-front orientation of time events. In Aymara, the word for past is *nayra*, which also means eye, sight, or front. The word for future is *qhipa*, also meaning behind or back. Tomorrow, *qhipüru* in Aymara, combines both day, *qhipüru*, and behind, *qhipa*, to form a word that in a literal translation would mean 'some day behind one's back'. Another example: the equivalent in Aymara of 'a long time ago' is *ancha nayra pachana*, in which *ancha* translates into 'a lot', and *pachana* into 'in/on/at time', hence, the word-by-word translation of *ancha nayra pachana* would be 'a lot of time in front of someone'.

As unique as these spatial relations might appear to us, they are very logical within the Aymara reality. Aymara people accord great importance to vision and to events that have been eye-witnessed. The language's grammar has a system of markers that differentiates whether the speaker was or not present when the action described in a sentence was realized. In every Aymara sentence, the speaker has to clarify the source of information, be it personal or impersonal, and both may be marked with verbal inflection or syntactic structures [113, p. 34]. This great reliance on vision to judge the veracity of what is uttered is reflected on Aymara speakers' preference to talk about the past instead of talking about the future. Speculating about the future, since one have never seen it, can come across as uninteresting, or even boastful. "As one Aymara said, 'The future in Aymara is what has not been seen. We cannot see the future... In Aymara the future is behind you—you cannot see it. In English the future is ahead of you, you can look into it' " [113, p. 33].

**Summary.** The 'elasticity' of the concept of time as shared by people in Latin America.

– $\Sigma$ –

Several publications, from business books to travel guides, usually emphasize the 'elasticity' of the concept of time as shared by people in Latin America. In her book 'The Culture Map: Breaking Through the Invisible Boundaries of Global Business', writer and professor Erin Meyer classifies cultures from places such as India, Africa, and South America as being time-flexible. Inhabitants of these regions have an altogether different level of elasticity in their minds with respect to time. Advice is offered to people traveling or doing business in these places to expect delays, since in South America, for instance, schedules and deadlines might not have the same importance as they would have in places such as Japan, North America and some parts of Europe – places she classifies as time-linear. The difference between linear and flexible time cultures lies in how tasks are approached and schedules are organized. Time-flexible cultures tend to deal with multiple tasks at the same time, valuing adaptability. Linear-time cultures, on the other hand, privilege sequential organization and focus on promptness.

Beyond business and corporate settings, there is a whole etiquette behind the ways in which South-Americans organize their time and how punctual they would be when they have to show up to an appointment. In Brazil, for example, it is expected that businesspersons and college professors be attentive to the time of their different appointments and classes, and any students or business partners showing up late would be probably frowned upon. Things become more flexible as people move from formal to more intimate scenarios. Actually, you are expected to arrive some minutes late at a dinner party at friend's house, for instance. It is all a matter of allowing your host to have some more time to finish cooking or putting on make-up before you show up to their house.

What seems to be just a carefree relationship with clocks and schedules may find its roots in the very way in which South-American societies were formed. Latin America had been for centuries under the influence of severe colonialism from Spain and Portugal before their countries could achieve independence. The imposition of agriculture-based economies before the first signs of industrialization were envisaged made the relationship to time evolve, in South America, in a much different way than it did in the United Kingdom, for instance. The time of sowing is different from the time of starting a machine. The sun's clock, which dictates the rhythm of the life in the countryside, is different from the wristwatch that commands the life in a European capital. The clocks that started to dictate the organization of public life in Europe in the Middle Ages were the perfect tool to organize the life of workers during the Industrial Revolution. Clocks did not work as successfully to manage the schedule of workers in the fields, to whom the rhythms of nature had a much greater importance.

The differentiation entailed by diverse economic needs is one of the factors causing a chasm in the experience of time between people of these cultures. In one side, there is the supremacy of the clock time and the careful management of schedules. In the other, the importance falls on engaging on activities at suitable moments. As mentioned above, anthropologist Thomas J. Bruneau coined the terms polychronic and monochronic time cultures to describe these two systems of viewing and experiencing time. Polychronic-time cultures, such as South America, value the involvement of people in activities rather than the adherence to schedules. References to events described by means of time often lack in precision and objectivity rather than pointing to a specific point in the time axis. A typical example is the use of the words *ahora* and *agora* (respectively the terms in Spanish and Portuguese for *now*). If a speaker of one these languages tells you that they are completing a task *agora* it does not necessarily mean that they are finishing the execution of the activity at the precise moment of the utterance, or as soon as possible. It can possibly mean that, but it can also mean that they will take care of the referred job maybe in a few minutes, or sometimes in a few hours. The diminutive forms *ahorita* or *agorinha* emphasize this hidden meaning: although the speech of your interlocutor signs that they will do something ‘now’, the real intention of the person uttering the word *ahora* is probably not to fulfill this promise, but to do it in some unclear point in the future. Monochronic cultures, on the other hand, stress the adherence to deadlines and the careful management of time. Both systems have their strengths and weaknesses and—help us understand that the relationship to time is a product of cultural influence.

**Impact.** It is remarkable that the cultures of time of North vs. Latin/South America are radically opposed. Whereas the United States cultivate an industrial absorption—and negation—of time within a profit-oriented set of ‘futuristic’ values, Latin America still cultivates the non-linear temporal practice, where the individual ontology of humans generates time culture.

– $\Theta$ –

5

# Genealogy and Ontology of Human Time Perception

**Summary.** This chapter summarizes two neurological aspects of time perception: in children and neurological localization on the one hand, including Parkinson and Alzheimer diseases in their influence on time perception, and also the impact of handedness on temporal processing on the other. We should also mention the fMRI-supported research [19] on neural emotional responses to musical performance and its expressivity. This research is "consistent with the idea that music performance evokes an emotional response through a form of empathy that is based, at least in part, on the perception of movement and on violations of pulse-based temporal expectancies." The movement-related perception emphasizes the temporal character of emotional response to music.

This short chapter is not a comprehensive overview of human time perception with respect to its theories in neurology and brain research. Our selection of a few relevant references among a large set of publications is driven by the background of this book's approach: contributions to our thesis that the reification of time perception is strongly correlated to gestural categories.

– $\Sigma$ –

## 5.1 Time Perception in Children

**Summary.** We summarize and discuss the important paper *Time perception in children: A neurodevelopmental approach* [36] by Sylvie Droit-Volet, which summarizes research on the development of neuronal time management in childrens' maturation process.

– $\Sigma$ –

Sylvie Droit-Volet summarizes:

> "In this review, we discuss behavioral studies on time perception in healthy children that suggest the existence of a primitive "sense" of time in infants as well as research that has revealed the changes in time


G. Mazzola et al., *Making Musical Time*, Computational Music Science,
https://doi.org/10.1007/978-3-030-85629-8_5

judgments that occur throughout childhood. Moreover, a distinction is made between implicit and explicit time judgments[1] in order to take account of the different types of temporal judgments that emerge across ages. Influenced by the philosophical conception of time, psychologists initially suggested that time is not a primitive dimension that can be directly perceived, but rather one which is inferred from the coordination of movements in space [127]. However, numerous studies have gathered data demonstrating that animals (rats or pigeons), which have no sophisticated reasoning capacities[2], are able to estimate temporal intervals accurately.

Since the publication of these results on animal timing, most researchers have come to agree that human beings and animals must all share a "sense" of time together with its hallmark characteristic: the scalar property[3]. The ability to estimate time is therefore thought to be universal. In this review, we will discuss behavioral studies on time perception in healthy children suggesting the existence of a primitive "sense" of time in infants, and those showing the changes in time judgments that occur throughout childhood.

Using a variety of paradigms, researchers have thus been able to demonstrate that infants are able to learn the temporal intervals between two events as well as the durations associated with events. (...) Infants are able to estimate the duration per se of an event. (...) This clearly demonstrates that the ability of infants to discriminate durations obeys Weber's law and is proportional rather than absolute. (...) Consequently, it seems reasonable to suppose that the cerebral mechanism underlying this sense of time matures early or/and is functional at an early age."

The discussion then tuns to neurological mechanisms for time judgment. The author concludes:

In conclusion, the prefrontal cortex, which has been shown to play a predominant role in cognitive time management, is therefore one of the late-maturing brain structures. The lack of maturation of the prefrontal and parietal cortices in children may therefore account for their diffi-

[1] An *implicit* time judgment is a non-overt estimate of stimulus or action duration. Timing emerges as a by-product of the temporal dynamic of regularly experienced perceptual or motor inputs. It is therefore an automatic processing of time permitting, for example, the anticipation of an upcoming stimulus without any conscious processing of time. In contrast, an *explicit* time judgment is an overt estimate of duration. Here, participants must consciously and deliberately estimate the duration of a new event.

[2] Not so evident: Pigeons may perform with great success logical operators, see [91, Ch. 6, p. 75].

[3] This relates to Weber-Fechner's law of human perception being proportional to the values, which is equivalent to the logarithmic perception, as known for pitch and loundess in music.

culties in attending to time or in maintaining temporal information in working memory.

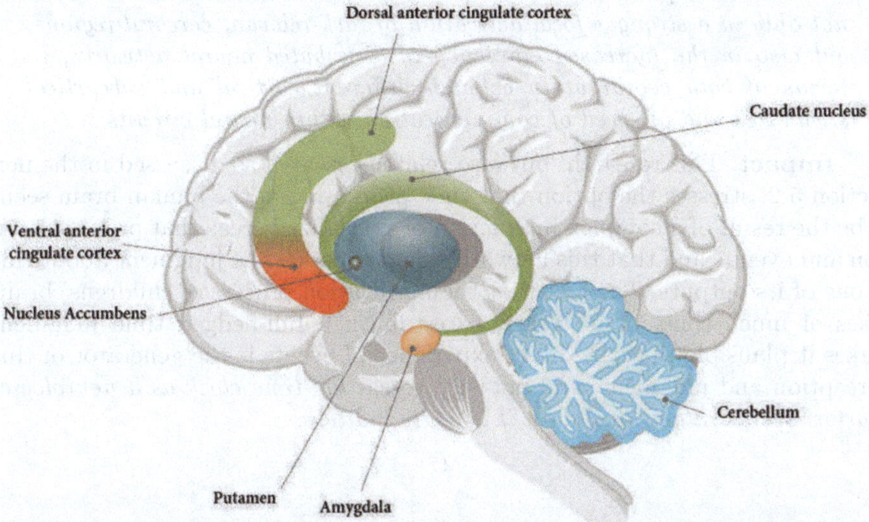


Fig. 5.1: The fronto-striatal system of the human brain may be involved in the perception of time.

> Overall, the data on different brain maturation events (myelination of axons, pruning and neural elimination) suggest that the functioning of the fronto-striatal system (see Figure 5.1) and its connections with other cerebral structures involved in the perception of time is less efficient in children. (...) In sum, the various fMRI studies indicate that the structural and functional maturation of the fronto-striatal system and its connections with other structures are associated with an age-related improvement in time discrimination and attention capacities. (...) The integrity of the fronto-striatal neural network is thus important for the different functions underlying time processing. As previously suggested, these regions have been shown to be structurally and functionally impaired in children with attention-deficit hyperactivity disorder (ADHD). In line with the hypothesis of a fronto-striatal dysfunction, ADHD children have been found to exhibit significant timing deficits in different types of temporal tasks.
> (Italics by this book's authors) *Overall, a variety of developmental studies provide a series of strong arguments in support of the important role of the structural and functional maturation of the fronto-striatal systems in the development of attention and executive functions underlying the explicit judgment of time. However, it is difficult to distinguish*

*between the processing of timing per se, on the one hand, and attention and working memory processes, on the other, because these are closely interconnected in explicit time judgments.*
*As neurodevelopmental studies have suggested, brain maturation results not only in a stronger focal activation of task-relevant cerebral regions, but also in the increased efficiency of distributed neural networks, in terms of both coordination between different cortical and subcortical structures and of speed of communication within neural circuits.*

**Impact.** This research, but also related research as discussed in the next Section 5.2, stresses the option that time processing in the human brain seems to be the result of a complex interaction of specialized areas that process attention and events, and that this network would generate the judgment about time as one of its outputs. The fact that the maturation process of childrens' brains takes of much time, many years, to establish a full-fledged time judgment, makes it plausible that the direct experience of events is the generator of time perception and judgment, and not vice versa: *the time clock as a neurological a priori seems rather secondary if it really matters.*

– $\Theta$ –

## 5.2 Neurological Localization of Time Processing

**Summary.** We discuss some representative recent neuro-scientific research results regarding the human brain's time processing; also referring to [39], [118], [166]. Our selection follows our main thesis about the genealogy of human time perception.

– $\Sigma$ –

Frasinatti et al. [39] state:

In the last two decades, the majority of investigations in the field of temporal cognition focused on the identification and the role of brain regions associated with time representation and processing. All these neuropsychological, neuroimaging, and brain stimulation studies in humans together with electrophysiological works in animals show that several brain areas, including basal ganglia, cerebellum, and posterior parietal, premotor, prefrontal, and insular cortices, are engaged in temporal computations. Although the role of each of these "putative" time regions is unknown, it is becoming increasingly clear that time in different scales and the use of time for different purposes are associated with different temporal networks and subserved by distinct temporal mechanisms.

This conclusion confirms the results by Droit-Volet regarding the non-localization of neural time management in children as cited in the previous section.

Timothy Muller and Anna C. Nobre in their paper [118] refer to different traditional time concepts and stress the non-genuine nature of time perception, its secondary role in human neuronal processing:

> Apparently dedicated timing mechanisms may in fact be incidental to the processing of other attributes. This distinction between dedicated versus intrinsic timing somewhat echoes the distinction between absolute, Newtonian time and relational, Leibnizian or Kantian time, respectively: time that is not extracted from its context is by definition not "in and of itself... without reference to anything external," but rather defined by the process performing the function in which time is intrinsically processed.(...)
> It is possible that the basal ganglia and cerebellum serve as centralized clocks, despite the evidence of localized timing mechanisms. However, it is not clear whether these timing mechanisms are dedicated to the problem of timing, or whether timing is a process inextricably wrapped up in other computational analyses being performed: for example, associative learning in the cerebellum or selection in the basal ganglia.

**Impact.** What is interesting in these papers is that the authors suggest that timing may be embedded as a consequence of other information, such as, typically, events, and not directly yielding a time dimension, but a network of events which envelop the category of time. This strongly supports this book's main thesis that time is a construction that unfolds from an interaction of gestures.

– $\Theta$ –

### 5.2.1 Pathological Temporal Aspects of Embodiment

**Summary.** We discuss the alteration of time perception for Parkinson and Alzheimer patients.

– $\Sigma$ –

#### 5.2.1.1 Time Conception in Music and Dance Therapy for Parkinson's Disease

**Summary.** In this chapter, music therapy and dance therapy will be explored from a musical perspective, especially rhythm, to understand the way it affects the Parkinson patients' neuronal chemistry in the perception of time. These findings will follow from possible connections between Parkinson's disease and music/dance therapy. Perception of time is important, and it is altered for Parkinson patients. These symptoms that can be ameliorated through dance/music therapy. In this chapter, we discuss the connection between the neurological rationales for Parkinson's disease and the successful application of dance/music therapy.

– $\Sigma$ –

**Introduction** Let us first give a short characterization of Parkinson's disease. It is a progressive nervous system disorder that affects the movement of daily activities. Parkinson's disease occurs when neuron populations produce less of the dopamine neurotransmitter, which causes the movement and time perception problems for Parkinson's patients. As a neurodegenerative movement disorder, Parkinson's disease is characterized by a constellation of motor symptoms that manifest tremors but also commonly causes stiffness or slowing of movement and leads to a significant decrease in quality of life. The alteration of time perception is manifest when Parkinson patients have to estimate time intervals. It turns out that they typically underestimate time interval durations [116]. This is remarkable since it relates time perception to movement processing in humans.

Second, let us give a short characterization of movement/dance therapy. The American Dance Therapy Association's official definition is this: Movement/dance therapy (MDT) deals with a therapy of human movements and dance with specific targets: MDT is defined as the "psychotherapeutic use of movement and dance to promote emotional, social, cognitive, and physical integration of the individual," to improve health and wellbeing. Despite this official setup, we should make more precise the logical context here: The general category of movement therapy includes movements with musical expressivity, which could be associated with dance. But there is also movement therapy without music, namely physiotherapy, which is a physical therapy for improving body injuries by appropriate movements without any musical context. Finally, the general category of music therapy, including also singing without movements, can be connected to movement therapy and thereby would be called dance therapy. This is the conceptualization we shall use in this book. The following Figure 5.2 visualizes this categorization:

Let us now discuss the connection between dance therapy and Parkinson's disease. We shall not discuss the connection between music therapy (singing) and Parkinson's disease, which has been investigated in [?].

Dance therapy has a connection with neurology, in fact, dance therapy, as a form of movement therapy, could particularly help with Parkinson's symptoms. We refer to publications of investigations of the effects of dance therapy on Parkison's patients [111]. Dance therapy was applied by a professional dance therapist to these patients, especially rhythmical dance akin to Tango. The effect of this therapy was measured by means of standardized criteria of the Movement Disorder Society Unified Parkinson's Disease Rating Scale (MDS UPDRS).

We conclude that the involvement of music in dance therapy offers an interesting perspective for future investigations on the pathology of Parkinson's effects on time perception, since time is a characteristic dimension in music, and even more so in its rhythmical structures.

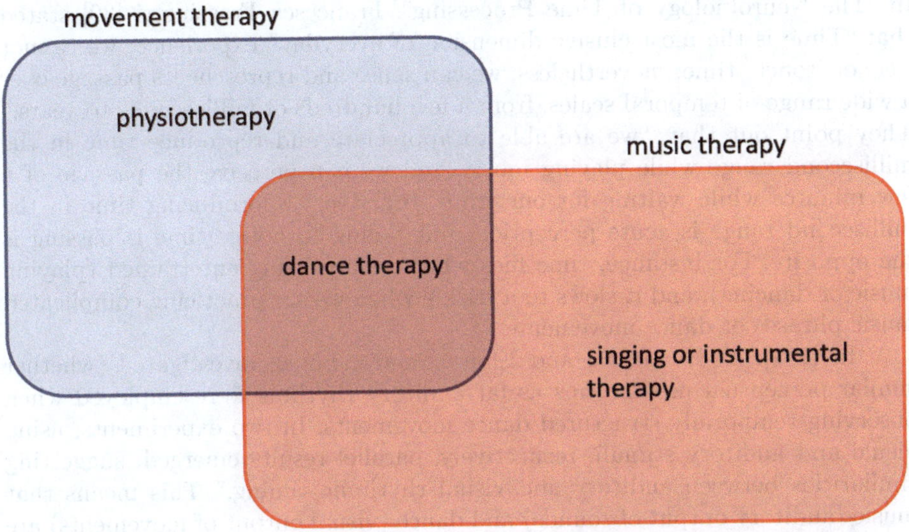


Fig. 5.2: The classification of therapies. Two top classes are movement and music therapies. Their intersection is dance therapy. The complement of dance therapy in movement therapy is physiotherapy, the complement of dance therapy in music therapy is singing or instrumental therapy.

How can we characterize dance therapy? Dance therapy not only speaks to the aesthetic, educational, and recreational parts of dance, but it also incorporates the bio-psychosocial (complex interaction of biological, psychological, and social factors been developed within an individual. A dance therapist focuses on movement behaviors as they emerge in the therapeutic relationship and environment (the therapeutic relationship between movements and emotions). The therapist uses movement to facilitate a compassionate relational opportunity within each session in a holistic body-centered approach to promoting health and wellness. In dance therapy, participants are encouraged to express emotions (Latin etymology: e-movere = to move out) through movement. Body awareness and mobility are addressed as in other approaches to dance, but with a unique emphasis on social support and emotional recognition. In this way, dance therapy combines traditional "talk therapy" with a body-centered approach to health and wellness. Human music, which is an essential component of dance, is always made by gestural/dance movements in the interaction with instruments. Therefore, the music dimension of dance and the dancing dimension of music are mutually interconnected (need more information to understand this from what has been written above). This is a significant result: the musical dimension of time influences human time perception.

**The Neurobiology of Time Processing**: why the dopamine processing is connected to time perception, and why dance is involved in this interaction.

In “The Neurobiology of Time Processing”, Francesca Frassinetti [39] stated that “Time is the most elusive dimension of everyday” experience; we cannot ’see’ or ’touch’ time; “nevertheless, we can sense and represent its passage over a wide range of temporal scales, from a few hundreds of milliseconds to years.” They point out that, “we are able to appreciate and reproduce time in the millisecond range while playing music, and we can perceive the passage of a few minutes while waiting for our train to arrive.” Reproducing time in the millisecond range is acute perception, but failing to notice time is passing is the opposite. For instance, time flies when we are being entertained (playing music or dancing), and it slows to a trickle when we are practicing complicated music phrases or dance movements.

In [153], Yi-Huang Su and Elvira Salazar-López investigated “whether similar perceptual mechanisms as far auditory rhythms were employed when observing temporally structured dance movements. In two experiments, using visual and auditory stimuli, respectively, parallel results emerged, suggesting similarities between auditory and visual rhythmic timing.” This means that music (auditory output of sounds) and dance (visual output of movements) are processed by the same perceptual mechanisms.

**Brain Chemistry Research** This section looks at brain chemistry and perception of time for healthy and Parkison’s disease individuals. In [166], they revealed “dissociable neural networks for the processing of duration with motor or perceptual components.” Their results suggest that the processing of temporal information is mediated by a distributed system that can be differentially engaged depending on the task requirements.

According to [150] they claim that “Midbrain dopamine neurons have been implicated in absolute and relative time processing. However, a direct link between signals carried out by dopamine transmitting neurons and temporal judgments is lacking.” They measured and manipulated the activity of dopamine transmitting neurons as mice reacted to the duration of time intervals. They found that

a) “pharmacogenetic suppression of dopamine neurons decreased behavioral sensitivity to time estimates,
b) Dopamine neurons encode information about trial-to-trial variability in time estimates,
c) transient activation or inhibition of dopamine neurons was sufficient to slow down or speed up time estimation, respectively.”

They suggest that “flexibility in time estimation may confer an adaptive advantage on the individual. For example, underestimating duration in better-than-expected situations may lead to longer engagement in those situations, resulting in even greater reward than if time estimation were not flexible.” Overall, dopamine neuron activity thus reflects and can directly control the judgment of time. This seems to explain why Parkinson patients suffer from an altered time processing. For this reason, time deficiency can be explained by the corresponding dopamine mechanisms.

To our knowledge, in the published literature, the connection between dance therapy and neurology has not been emphasized so far. Previous studies in dance therapy support that dance can benefit Parkinson patients, although the music elements in dance therapy haven't been explicitly discussed. On the other hand, we don't know of any such studies looking at dance therapy for its connections to neurological processes. But there are studies in the dance therapy discipline, which focus on its benefit to patients' conception of time and how it affects patients' bodily, emotional, cognitive, and social integration and wellbeing.

*The logical conclusion is that since dopamine and time perception as well as time perception and dance therapy are connected, the connection of dopamine and dance therapy is also a fact.*

*Dance therapy* activity requires planning movements, following music and signals, and remembering choreography. These music/dance (dancing and listening to music with other people) exercises improve patients' conception/sense of time via physical body movements. Since dance therapy helps patients with Parkinson's disease in their movements, we conclude with the following hypothesis:

*The Parkinson patients' bodily, emotional, cognitive, and social integration and wellbeing generated by dance therapy influences their neurological dopamine processing.*

#### 5.2.1.2 Alzheimer: Two Times

**Summary.** Alzheimer patients' memory performance via musical triggering [77] and time distortions [48]. Also refer to [30].

– $\Sigma$ –

There is strong empirical evidence of Alzheimer patients showing a distortion of time perception [48]. This is more precisely manifest when these patients are given verbal instructions and then tested about their time judgments.

For example, such tests asked participants with Alzheimer to read either 5, 10, 20, or 40 digits appearing one at a time. At the end of each sequence, participants had to judge the elapsed time. Results showed inaccurate time estimations in Alzheimer participants, especially for long intervals.

The most distinguishing characteristic of human memory is mental time travel, or the state of autonoetic consciousness permitting the episodic reliving of past experiences. By inducing the feeling of subjective time, mental time travel is likely to depend on time perception, which may explain why, in the work of El Haj et al. 'Remember' responses, reflecting autonoetic reliving, were significantly correlated with time perception. It is worth noting that, in the same study, no significant correlations were found between time perception and 'Know' responses, reflecting a dissociation between time perception and noetic consciousness. Time distortions may impair mental time travel, and consequently may contribute to an episodic memory impairment in Alzheimer.

El Haj speculates that, in Alzheimer, this reliability is degraded due to memory impairment potentially resulting in time distortions and longer decision times, collectively aggravating patients' feeling of being 'stuck in time'.

The time concept that was studied in these investigations was however not a musical one. The investigations of Manuela Kerer et al. [77] focused on mild cognitive impairment (MCI) and early-stage Alzheimer (AD) patients, who were compared to healthy individuals' response to musical tests "in terms of verbal memory of music by the identification of familiar music excerpts and the discrimination of distortion and original timbre of musical excerpts."

The results showed that "MCI and Alzheimer's patients showed significantly poorer performances in tasks requiring verbal memory of musical excerpts than the healthy participants. For discrimination of musical excerpts, MCI and AD patients surprisingly performed significantly better than the healthy comparison subjects."

In conclusion, the paper supports "the notion of a specialized memory system for music."

This study strongly focused on melodic structures, which are strongly dependent on the two-dimensional structure of melodies in the time and pitch domain. This implies that the perception of musical time must differ from the perception of non-musical time that was tested in the previous investigations as referred to in [48].

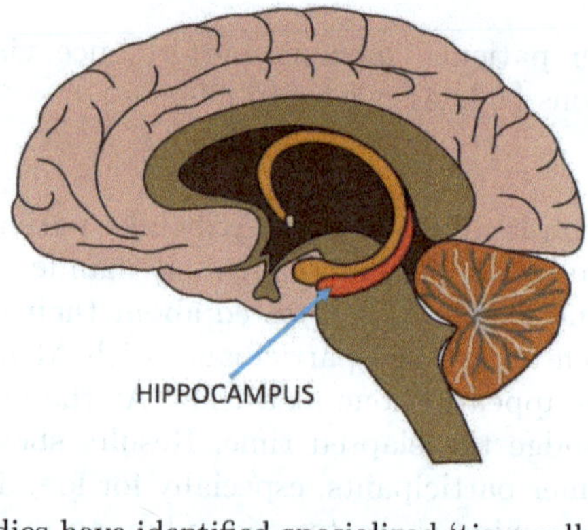


Fig. 5.3: Research studies have identified specialized 'time cells' in the hippocampus.

In [48], it is argued that "neuropsychological research suggests that individuals with damage to the prefrontal cortex have difficulties to project themselves in past or future. More precisely, timing difficulties and memory compromise can be related with compromise in the hippocampus." It is interesting that the projection to past and future is a time-feature that philosophers classified as a detached experience, especially Merleau-Ponty (Section 2.10) would have disqualified this temporal feature from existential time reality. This reference to the hippocampus seems a delicate conclusion since our own studies of hippocampal Depth-EEG structure stemming from the perception of consonant and dissonant musical intervals [94, Ch 30] suggest that the hippocampus is a musically sensitive part of the brain. If there is a specialized system for music in the human brain, and if this system still works for AD patients, the hippocampus must have components which remain healthy for music, and especially for musical time perception. We should recall here that the hippocampus is a bipartite organ: left and right hippocampus may work differently, a property that became evident from EEG profiles found in the investigations [94, Ch 30].

*These investigations suggest that musical time is a time category that differs from non-musical time in humans.*

Recent investiations [30] confirm these findings. Here is the paper's abstract:

> A later symptom of dementia is a decline in expressive language. Together with memory loss, this can create major communication challenges and high levels of frustration. Research into the use of music as a therapeutic intervention for dementia patients has shown that, even in the late stages of the disease, there remains a capacity to enjoy and respond to music and to learn and recall new music. This sparing of musical ability, together with the fact that language is more readily recalled when sung rather than spoken raises the possibility that a music intervention using a mnemonic strategy may maintain some level of verbal communication in people with dementia.
> This was the focus of a six-month pilot study. Once-weekly music sessions were delivered to residents of two care homes specializing in the care of people with dementia. Each session was in two parts. A sing-along that incorporated songs familiar from their teenage years primed musical engagement. This was followed by the introduction of 'micro-songs'-functional language relevant to daily life set to short specially composed rhythmic melodies. In addition to increased confidence and social interaction, results showed that participants learned the new materials in an average of four weeks and could recall them after intervals of between one week and two-months. Cognitive ability was not a determinant of their capacity to learn and recall the resources. Future research will build on these findings and investigate whether these resources can be embedded in daily life and used to facilitate communication.

### 5.2.2 Psychology of Metrical Perception

We cannot delve into the complex psychology of time perception, especially of meters and rhythms, but refer to Justin London's in-depth discussion of such topics in his book *Hearing in Time* [80]. As a remarkable observation, he cites Bruno Repp's insight [138, p. 978] that "rhythm perception may simply be internalized sensory motor synchronization."

**Impact.** The Parkinson patients' bodily, emotional, cognitive, and social integration and wellbeing generated by dance therapy influences their neurological dopamine processing. The investigations regarding Alzheimer suggest that musical time is a time category that differs from non-musical time in humans.

– $\Theta$ –

## 5.3 Handedness

**Summary.** This essay proposes that a relationship between a particular variant of handedness and creative instrumental performance yields unusual expressions of musical rhythm and time. Alex Lubet describes here "left-centric performance"—that of a strongly left-handed person playing string instruments right-handed—leading to a highly personalized approach to rhythm and time. It is characterized at nearly all times by a greater proportion of lighter markings of time than are normative in right-centric performance. Left-centric time marking is characterized by soft articulation, including "softer" musical features, ranging from prevalence of legato, textures dominated by overlapping sustained tones of long but uncontrolled duration, and tones with inaudible onsets, the latter activated without touching the strings at all, but with a distinctive role for the left hand.

Lubet's intention here is to propose the idea of "left-centric, right-handed (creative) performance" of string instruments, with an emphasis on the creation of rhythm and time. That Lubet has offered only examples from his own compositions and improvisation is not intended as a claim to uniqueness, but as an invitation to left-handed, right-playing colleagues to reflect upon their playing and composition, consider whether they share Lubet's particular handedness orientation, and, if they find common ground, share experiences and musical examples.

– $\Sigma$ –

### 5.3.1 Handedness in Culture and Science

Interest in handedness—a preference for the use of the right or left hand—is evidenced as early as the ancient Assyrians. A ubiquitous expressed bias toward right-handedness across time and throughout cultures, appears to have always been prevalent at a rate of about 90% (slightly higher in males than females), in all major ancient civilizations, and is found in such thinkers as Plato and Aristotle. The Hebrew and Christian Scriptures reflect the same bias (Right, Left, Right, Wrong).

The preoccupation with handedness, often manifested in scientific studies, has persisted to this day. As might be expected, handedness studies are often particularly concerned with the exception, left-handedness. (Not surprisingly, many scholars of left-handedness, and occasionally mixed-handedness (Christman; Hendrix), identify as "lefties," Lubet included.) This situation is to a degree analogous to the way in which most studies of race, gender, and sexuality focus upon marginalized or minoritized groups and that scholars from these groups are prominent in these fields. One difference between handedness studies and those of the other groups mentioned above is that the former more likely to appear in the "hard" and behavioral sciences, notably psychology, than in the humanities and social sciences.

While prejudice is a frequent topic in studies of left-handedness, it is a lesser concern than in race, gender, sexuality (and intersectional) studies. An additional difference, as evidenced in my review of the literature, a sampling of heavily cited works, with a bias toward music, makes no reference to left-handedness as a social construction. Throughout the literature, handedness is presumed to be an objective reality, albeit without a universally accepted definition. Further, many authors recognize that right-handedness has often been a forced choice, grounded in societal norms. Left-handedness is also occasionally a forced choice, the consequence of an injury [129, p. 12].

### 5.3.2 Handedness in Music

While handedness is often identified according to a single activity—writing —a more sophisticated understanding that is associated with a more sophisticated classificatory instrument such as the Edinburgh Handedness Inventory (EHI) [121]. The EHI determines handedness according to hand use preference for a variety of activities (ten in the original study). While playing music is not among the these, the inventory was developed "In the course of a search for left-handed musicians." [121, p. 99] Some studies that employ the Inventory have found a greater representation of left-handedness among musicians than in the general population (for example [3], [70], [131]). Christman [25] notes that "mixed-handedness is over-represented among players of string instruments." Statistical studies and empirical writings detect advantages [33], disadvantages, and neither advantage nor disadvantage [71], [128], [45] among left-handed musicians. [26] associates left-handed guitarist Jimi Hendrix's exceptionality as both a guitarist and composer/lyricist to his overall mixed-handedness.

Lubet's particular interest here is in those stringed instruments where the hands (for the most part) function in distinct roles; one (primarily) to activate the strings, with a bow, plectrum, or fingers—that is, the rhythm hand—and the other (mostly) to control pitch by depressing the strings against the (fretted or non-fretted) fingerboard. Universally, the normative division of labor is principally *right hand for rhythm* and *left hand for pitch.* By contrast, keyboards, percussion, and woodwinds have more similar roles for the hands, while brass instruments are strongly one-handed; left for the French horn (although the right hand had the more difficult role of controlling pitch prior to the introduction of valves), right for all others, with one hand working the valves or slide.

Thus, a discussion of handedness in instrumental performance will yield the richest information in the context of those string instruments with different tasks assigned to the hands, such as the violin and guitar families (henceforth "string instruments"), as contrasted, for example, with various harps and hammered dulcimers. *In this volume, it is of course of paramount interest that the division of tasks between the hands is in large part a division of pitch versus rhythm.* This is a generalization, but this broad characterization is nonetheless significant.

It requires only empirical evidence to observe the culturally imposed challenges to left-handed string players. For guitarists (and performers on similarly plucked/picked instruments), this may mean the formidable difficulty of obtaining a limited selection of left-handed instruments, as played by Paul McCartney, playing a right-handed instrument upside down with (like Jimi Hendrix) or without (like Elizabeth Cotten) reversing the strings, or playing right-handed (Mark Knopfler). For performers on the bowed string instruments of the symphony orchestra, there are the additional difficulties of the orchestral seating regimen (in which all players in a section face the same way) and the resistance of teachers to accepting students who play left-handed [129, p. 48].

Because of the different tasks assigned the hands in playing string instruments, there are differing opinions as to which hand does the more difficult work. It is often opined that the fine nuances of pitch control make the left-hand role more challenging ([46, p. 44], Thomson (2003), quoted in [129, p. 11], [26, p. 260]). Were this actually the case, then the universal rendering of the assignment of the rhythm task to the typically dominant right hand would be a mystery.

Despite frequent arguments to the contrary, there is a clear consensus that, for instruments played primarily with a bow, the bow hand, virtually always the right, has the more difficult task; that is, it requires more fine motor control, because the bow hand does not directly touch the strings, as does the left [129, p. 11].

The question of which hand role, pitch or rhythm, poses the greater challenge is more complex in the case of guitar and related instruments, because of both the variety of instruments and the range of difficulty of dominant (typically right) hand playing techniques, from strumming to fingerpicking to use of one or more plectrums. In *The Puzzle of Left-handedness*, Smits discusses this at length [147, pp. 90-92]. He acknowledges that the typical left-hand task of controlling pitch *appears* more difficult, perhaps especially in the violin family where there are no frets and a large expressive role for the frequency modulation known as vibrato. He nonetheless concludes that the job of the right hand, even when strumming, presents the greater challenge [147, p. 92]:

> The fact that in the case of the violin and guitar we always leave the job of fingering to the non-favoured hand shows that in our apparently simple strumming and bowing lies something that makes considerable demands. *It must have to do with rhythm* [emphasis Lubet], and with

having complete mastery of the dynamic. It's not a matter of strength but of control, of minimal, extremely precise differences in the force, pressure, and speed of movement. Strumming and bowing involve exceedingly subtle steering.

### 5.3.3 Left-Handedness in Creative Versus Rote Performance

Issues of rhythm in left-handed string playing would be of lesser interest were the topic limited to Western classical music and other repertoires in which the performer executes a pre-notated part that, by virtue of its tacit non-consideration of the handedness of no player in particular—even when the work is commissioned by a specific artist—implicitly assumes the default of right-handed performers. In a related field of music research, performing arts medicine, it has been observed by Manchester and Brandfonbrener that the vast majority of studies have used as subjects Western classical musicians. They attribute this both to the relative ease of recruiting performers from such stable institutions as symphony orchestras and schools of music and to the affinity of privileged researchers—that is, physicians and other well-schooled health professionals—for this repertoire. It understandably follows then that scholarly studies of left-handed string players—who are part of the larger class of disadvantaged, if not bedeviled or even disabled, sinistrals, mostly concern classical violinists [79], [129], [149].

Among left-handed classical violinists, nearly always compelled to play fastidiously notated and largely standardized, "common practice" repertoire right-handed, the question of whether handedness matters to creative performance—in musics where performers have significant responsibility for inventing their respective parts—or in the compositions of composer-performers remains unanswered. If, in classical music, there are interpretive differences related to handedness, including rhythmic differences, these have not been studied. Those few violinists actually able to perform on a left-handed instrument, whether converted or specially made, presumably play much the same as their right-handed colleagues, with their dominant and non-dominant hands executing their intended tasks. Lerner notes that left-handed folk fiddlers, who typically have more creative control over their parts than classical violinists, are also more common. Neither Lerner nor Pyykönen, leading authors on the subject of left-handed violin, refer to what Gottlieb [46] calls "idiodextrous" performers who play "upside down," that is, left-handed on right-handed violins with unreversed strings, though [149] gives the example of 18th century English child prodigy William Crotch (1885-1847). As a string player, the multi-talented Crotch was principally an idiodextrous cellist, who also played an idiodextrous violin, held like a cello. Guitarists Elizabeth Cotten, Freddie King, and Dick Dale, and electric bassist Jimmy Haslip are among a surprisingly large number of idiodextrous creative players and composers. At 900+ pages, John Engel's *Uncommon Sound: The Left-Handed Guitar Players Who Changed Music*

[37] is stunningly comprehensive and includes artists on both idiodextrous and reversed-string instruments.

Gottlieb, an idiodextrous jazz double bassist, has authored an exhaustive study of this phenomenon, the richest by far for any instrument. It includes discussion of five additional idiodextrous jazz bassists—arguably the only ones of significance—as well as four bassists who play left-handed instruments. One of his most important observations as concerns idiodextrosity in jazz strongly parallels what Lubet has written about jazz and disability [83, pp. 42-68]:

> The malleability of identities and cultures in jazz is arguably more receptive to unusual handedness approaches than is the classical tradition. In the context of jazz improvisation, idiodexterity may even facilitate the development of a 'uniquely idiodextrous' vocabulary of musical phrases and techniques [83, p. 2].

Idiodextrosity is the approach to left-handed string performance (and left-hander's choice of instrument) that most obviously affects the creative decisions of performers and composer-performers. What Gottlieb observes in jazz is similarly applicable to folk (Elizabeth Cotten), blues (Freddie King, Albert King, Cedell Davis, and Otis Rush) and rock (Dick Dale). Although he reversed the strings on his guitars (which requires additional mechanical adjustments), Jimi Hendrix's use of right-handed electric instruments, whose tone controls were above rather than below the strings—and thus more readily accessible *while playing*—had an important impact on his playing [26]. Although one of the best studies of left-handed string performance, Christman does not discuss the rhythmic aspect of Hendrix's playing at all.

Although there are certainly rhythmic ramifications for both idiodextrous performance and the use of "upside-down" right-handed electric guitars by Jimi Hendrix, either approach—as well as the use of standard right-handed instruments by artists such as Paul McCartney—enable the dominant hand—in these instances, the left—to have the principal responsibility for activating the strings and thus primary responsibility for rhythm. While the rhythm/pitch division of labor between the hands is a generalization and said division is a matter of degree that varies between instruments, styles, and method of activation—bow, fingers, plectrum(s)—it is nonetheless a useful distinction.

Our focus here though is a variant in handedness that most certainly does impact rhythm in creative performance and composer-performance. It originates in my research in disability studies in music, in which Lubet first proposed the idea of left-handedness as a disability in musical contexts [82], as well as his own string instrument performance and composition over more than five decades.

Many highly accomplished guitarists claim a left-handed identity while playing in a conventional right-handed manner. These include Duane Allman, David Bowie, Billy Corgan, Elvis Costello, Robert Fripp, Mark Knopfler, Johnny Winter, Joe Perry, and Paul Simon (Ultimate Classic Rock 2015). Lubet uses the qualifier "left-handed identity" because these artists' handedness has not been confirmed by a scientific instrument such as the Edinburgh Handedness Inventory (EHI). Handedness is not binary but, on a continuum, that includes "strong" and "weak" left and right, as well as a central category of mixed-handedness, the latter which includes the most celebrated left-handed guitarist of all time, Jimi Hendrix [26]. Oldfield [121, p. 98], inventor of the EHI, also notes the importance of "the partial adaptation of left-handed people to the constraints of right-handed implements," a classification that certainly, intended or not, includes string instruments.

Given the virtuosity of some of the artists referenced above, it is reasonable to assume that they may not be as strongly left-handed as, for example, Paul McCartney or Kurt Cobain, both left-handed players, or that they might be mixed-handed like Jimi Hendrix. Christman [25] demonstrates a greater prevalence of left- or mixed-handedness in string and woodwind players, who have greater bi-manual coordination, than performers on more hands-independent keyboard instruments. The degree of right-handedness exercised by anyone as a consequence of manifold forced choices is an important question, though beyond the purview of this essay. It seems reasonable that the greater bi-manual coordination found by Christman may derive to some degree from the need for left-handers to execute tasks right-handed on a consistent basis.

### 5.3.4 Left-Handers Who Play Right-Handed

Left-handed Megadeth guitarist Kiko Loureiro is surely not alone in having learned to play right-handed on a sibling's (or other only available) instrument. It is the reason Lubet plays right-handed as well, having continued that practice as he learned and acquired several other instruments. Young beginners may well be unaware of the availability of left-handed instruments. Although Lubet has memories of having picked up a guitar and holding it left-handed before he ever started playing, Lubet was never presented with the possibility of that alternative. His few years of lessons precluded any option other than right-handed play.

While Gottlieb centers his survey of idiodextrous double bassists with a brief reference to his own practice, Lubet finds himself required to limit his study to an $n = 1$, consisting of himself, while positing that this essay is only a beginning. Lubet classifies himself as strongly left-handed, rarely doing anything right-handed unless forced by the available equipment. (The only exceptions are those rare occasions when Lubet has swung a baseball bat or a golf club. That Oldfield observed that raking and shoveling are activities that right-handers often perform left-handed, seems to imply that reversing hands on similar sports equipment is common.) Lubet's right-hand facility has always been limited—Lubet would say seriously so—on all his instruments, the strings Lubet plays and composes for professionally, as well as piano, his principal instrument in university, see Figure 5.4.

No amount of practice has ever enabled Lubet to transcend his right-hand limitations. In an experiment using a non-music task, $n = 14$, Peters

Fig. 5.4: Alex Lubet playing the mountain dulcimer.

[126] demonstrated that "preferred hand superiority on a finger tapping task persisted in 13 out of 14 subjects even after prolonged practice," an observation that may have implications for musical performance, although some musical handedness studies suggest otherwise [71], [69], [70].

As is often the case in discourse concerning minorities, there is a tendency toward triumphalism, the celebration of those who have "overcome" disadvantage. Such is the case with left-handed "guitar heroes," regardless of the manner in which they play. A Google search for "left handed guitarists" yields an overwhelming majority of top sites that are concerned with the famous and successful. The responses to surveys conducted by `lefthandedchildren.org`, which surely reflect the experiences of greater numbers than do the predominant narratives of success and fame, tell a quite different story, summarized by [46, p. 44]:

> The attitudes range from those who didn't know there was a left-handed way to play, to those who 'gave up' (because the practical challenges of finding an instrument and a teacher made it 'too hard' or 'expensive'); from those who preferred to play right-handed and learned that way, to those who learned left-handed—usually with a fair amount of self-teaching involved.

Lubet's own experience is as a left-handed person who persisted in playing right-handed despite great difficulty and serious technical limitations while managing to have a lifelong career in music. Thus, it differs from the narratives of both the guitar hero and the frustrated child. Lubet's "handedness story" (another useful term coined by Gottlieb) is like those referenced above in that his learning was largely auto-didactic. Lubet avoided or quit the most challenging right-hand idioms, including classical guitar and double bass and mandolin in any genre. Lubet also quit using a pick, despite only having been taught that way. Most important, Lubet gravitated strongly toward musics in which Lubet either made up his own part, improvised (in a left-centric style) or bore full responsibility as composer.

As referenced above, Smits is one of several authors who disputes the common but mistaken—if not oxymoronic—misconception that the left hand has the more difficult task in right-handed string instrument performance. Although this is an incorrect notion it nonetheless carries with it the implication that left-handers who play right-handed may be naturally highly facile in left-handed playing duties. Lubet includes himself in this category and regard it as foundational to both his playing and his compositions. In particular, what might be regarded as "left-centric right-handed playing" (henceforth, simply "left-centric") has had a particularly profound impact on rhythm and on time concept writ large in his performance and composition.

Aspects of left-centric playing can be heard in Lubet's Aria (After Puccini) [84]), a solo acoustic guitar work based on Puccini's *O Mio Babbino Caro* from the opera *Gianni Schicchi*. The midsection of the performance (1:15-3:38) is an improvisation that features many fast runs, whose velocity owe exclusively to the use of two types of left-hand rhythmic activation of the strings: "hammer-ons" and "pull-offs" (Seeger, *How to Play the 5-String Banjo*, Folkways Records 1954).

Many styles of string playing employ these techniques, regardless of the handedness of the performer or the configuration of the strings. But these and other left-hand functions are exclusively the means Lubet uses to play fast and they guarantee a large measure of legato and consistently soft articulation and flow. There is no fast, single-note plucking or "tapping," as heard in, for example, jazz, bluegrass, or the "shredding" style in rock, heavy metal in particular (nor can Lubet play as fast as "shredders" (for example, "Mind blowing Female Guitarists SHREDDING!" Best in the world 2 [85]).

The same approach is heard in *Macedonian Dream*, a duet with its composer/arranger, (right-handed) classical guitarist Maja Radovanlija. It is helpful to hear the difference between their left-centric and right-centric improvisations. At 4:26 they alternate playing a particular variety of unmeasured tremolo, a right-hand, two-finger technique associated with—and easier and less painful on—the nylon strings Maja plays. It may actually be easier to create this effect with one's non-dominant hand, with less fine motor ability, which thus operates well in this context as a "blunt instrument." Lubet also uses this effect on (steel-stringed) electric bass and mountain dulcimer, where it is unusual, as well as (more typically) nylon-stringed ukulele.

While bending strings with the left-hand to alter pitch is an essential element of blues and rock guitar, Lubet's performance practice includes a particularly left-centric approach. The tempo of *Reliquary Dances I* [86] is very slow, so the pitch accuracy of bends is of utmost important. Particularly notable are upward bends of a Major 2nd (minor 2nds are easier and more characteristic) from the natural 7th scale degree to tonic (0:02-0:05), which demands utmost accuracy and considerable left-hand strength, when played on an acoustic guitar with heavier strings and higher action than most electric guitars, where bending is almost effortless. Even more demanding are passages where the string is bent silently to a precise pitch and slowly lowered (0:16-1:18). The degree of flow and the paucity of individually articulated pitches or chords (until 1:43), is, at this slow tempo, where rhythm is an even more left-centric function than in the fast tempo examples above, and where glissando, rather than legato, dominates. The final tonic note (2:05-2:13) is bent with even less articulation and greater flow, by means of moving the guitar towards and away from the microphones to create a Doppler shift [87].

The most left-centric of techniques frequently employed in Lubet's performance and composition practice—and that with the greatest influence on time-making is the extensive use of natural harmonics. High harmonics require a very light left-hand touch and, on the lower strings, Lubet is able to play these and use them in compositions up to the 16th partial (for example, *Eight Ouds No. 1*, [88].

In addition, Lubet is able to play single-string multiphonics, typically by plucking strings simultaneously with both hands, with the left hand on a harmonic node, such that both the natural harmonic and the open string are sounded. It is also possible to sound multiple harmonics on a single string by touching the string in between nodes or upon multiple nodes that are especially close together, with or without also plucking the open string. Typically two, but sometimes three, harmonics are possible on a single string. This yields the possibility of playing more pitches simultaneously than there are strings on guitar and other instruments. Multiphonics require an even lighter left-hand touch than individual harmonics and, because harmonic nodes often do not align with the frets, a great deal of left-hand accuracy. The highest harmonics are at least as close together as the highest notes on the violin. One compositional opportunity afforded by natural harmonics over a wide range of partials, as opposed to the artificial harmonics favored by performers on bowed strings (5th partial) and, to a somewhat lesser extent, classical guitarists (octave), is that different partials on different strings yield many different timbres.

Unlike stopped notes on string instruments, natural harmonics, like open strings, sustain while no longer being touched, and it is most idiomatic to permit these to ring as long as possible. Durations and dynamics are both, to a great degree, functions of the open string and the partial; the lower the string and the lower the partial, the louder and longer a harmonic or multiphonic may be sustained. Because of both the aesthetic desirability in composition and performance to allow these notes to sustain as long as possible and the variable sustainability of partials, according to their distance from their fundamentals/open strings, *musical time in such left-centric music is exceptionally closely related to the acoustical properties of individual sounds.* Often, there is little point in specifying dynamics or durations (other than their attack) of certain harmonics or multiphonics, as they can only be played one way.

Such individual sounds, unusual and often complex in timbre not only demand aesthetically that they be heard for their full possible value, they require syntactically that they be placed at the ends of phrases, where they can be heard to best effect. It is also often the case that the acoustical richness of such sounds, when they end phrases, as they so often do, demands that they be followed by a similarly long silence that permits reflection upon the complex timbre that has just been heard. The sequence of complex timbre—acoustically rich—and reflexive silence—acoustically empty—may be construed as a kind of cadence formula, a formal signifier, albeit one very different than the formulaic harmonic resolutions typical of Western music.

The above description evokes both the concept of *ma*, or interval/space, in Japanese arts, and composer/ethnomusicologist Chou Wen-Chung's observations about a key difference between Western and Asian musics. Chou proposed that the smallest unit of expression in Western music is the interval—and thus the harmonic cadence—versus the individual note/sound in Asian music—the cadence formed of the acoustically complex moment followed—indeed protected from intervallic syntax—by a temporally generous silence.

In *Reliquary Dances IX*, the opening (0:00-2:05) and closing (3:25-4:26) sections are exclusively in harmonics and multiphonics. The latter are first heard at 00:23, first in chords, at 0:40 also on single strings. The middle section, beginning at 2:06, emphasizes the kind of left-centric legato described above, with the addition of harmonics at 2:54.

Of particular interest is the final chord, beginning at 4:04. Five strings are played with 3 consecutive upper partials each (5th through 7th), plus open strings, yielding a 20-note chord, that is bent via Doppler shift at its close, and allowed to hold for its full resonant value. At this writing, Lubet has found only one location on the strings where three audible partials are close enough to cover with a fingertip and is only able to play five strings with multiphonics simultaneously. However, the acoustics of the guitar and the range of left-centric technique are a work in progress.

Another acoustically-driven, left-centric technique exploits partial relations in a very different way. In *Reliquary Dances VI* [89], beginning at 0:31, sympathetic open strings create the background harmony. Rather than being "played," they are activated when the notes of the pentatonic melody are their low partials. For example, the open D is activated by melody notes D, F♯, or A. This technique differs from the use of exclusively sympathetic strings of East Indian instruments such as the sitar and sarod in that the same strings alternate melodic and sympathetic roles.

What makes this a left-centric technique it that it is critical that the guitarist avoid touching and thus damping any string other that upon which melody is played at the time, such that the remaining open strings be permitted to vibrate sympathetically. This requires very facile left-hand technique and Lubet is unaware of other guitarists having exploited it this systematically. The open strings initiate without an audible attack and with remarkable sustaining power—when they are reactivated yet not rearticulated by successive melody tones—and a surprisingly large dynamic range, in correspondence with the dynamics of the melody tones, as well as their harmonic relationship. For example, when a higher partial of an open string is played in the melody, the open string's dynamic will be lower than if an octave above that string were sounded.

Sympathetic string harmony—perhaps "sustained tone accompaniment" is more accurate, if less elegant—in this composition operates somewhat differently than in classical Indian music. Unlike sitar or sarod, the guitar has no designated sympathetic strings. The tonal center and scale in this piece were chosen to cause the (mostly lower) strings of a guitar in standard tuning to resonate when they are not playing melody. That harmony is not "played"—the onset of the sympathetic pitch is unheard, covered by the louder melody tone that activates it—and can sustained at very great and somewhat aleatoric duration—the length of these notes is literally out of the guitarist's hands—creates a sense of time flow that is unusual in Western music for acoustic instruments in any genre.

Simply put, a plucked string music that so prioritizes timbre and demands that complex and sustainable timbres be permitted to resonate for their full durations is *a music whose time is shaped by acoustical properties.* It tends by nature to favor slow tempos and even slower moments and often prefers to set time according to the lifespans of sounds rather than by a regularly scheduled pulse. Further, a left-centric yet right-handed music that allocates a larger-than-typical rhythmic role to the left hand—theoretically non-dominant, but in practice more responsible for both speed and nuance—will mostly mark time with softer articulations and an outsize role for legato.

### 5.3.5 Conclusions

There are, of course, many left-handed, right-playing string instrumentalists who have well-developed right-hand technique, presumably acquired with only limited difficulty or none at all. These likely include the well-known artists mentioned above. It is also likely that right-handed, right-playing guitarists could emulate the most challenging of Lubet's guitar techniques, although he has seen right-handers struggle without success to play the high harmonics.

Perhaps the most certain conclusion one can make is that the unusual guitar techniques Lubet has developed originated in his left-handedness and occurred to him because of it. Lubet drew similar conclusions about the playing styles of disabled jazz virtuoso Django Reinhardt (whose style is much emulated by "Gypsy jazz" guitarists, some of who go so far as to play using only two left-hand fingers) and pianist Horace Parlan [83]. In the context of this volume, the music Lubet composes and improvises can be understood as the result of giving the left-hand extraordinary responsibility for the construction of musical time, a kind of time that is articulated more softly and sometimes even not at all.

**Impact.** Musical time in this left-centric music is exceptionally closely related to the acoustical properties of individual sounds the music. Lubet's composition and improvisation can be understood as the result of giving the left-hand extraordinary responsibility for the construction of musical time, a kind of time that is articulated more softly and sometimes even not at all.

– $\Theta$ –

# Part III

# Musical Time Concepts

# 6

# Meters and Rhythm

**Summary.** This chapter reviews first the socio-biological genealogy of rhythm and then Mazzola's mathematical theory of meter and rhythm, together with associated topologies, as described in his book *The Topos of Music.*

We refer to Roger Matthew Grant's excellent book *Beating Time & Measuring Music in the Early Modern Era* [52] for a historical discussion of musical meter theory.

$-\Sigma-$

## 6.1 The Genealogy of Meter and Rhythm

**Summary.** This section reviews Mazzola's mathematical theory of meter and rhythm, together with associated topologies (as described in his book *The Topos of Music* [94, Vol. I, Ch. 21]). Metrics and rhythms are excellent elementary test-cases for global structures in music. We shall critically review two common approaches, Riemann's and the Jackendoff-Lerdahl theories. We then develop the concepts of global time structures and their topologies, including associated weight functions.

We refer to the introductory discussions of local meters and global meters. We shall above all concentrate on the metrical aspect since it yields a simple and formally transparent situation from which the rhythmical complexification can easily be unfolded.

$-\Sigma-$

### 6.1.1 Biological Genealogy

Rhythm is ingrained in our daily life; it affects our moods and feelings. Music is developed not just as an intellectual magic, it is heavily connected with our embodied flow. We breathe at a certain speed, with periodic rhythmic patterns. With our resting heart rate and breathing rate, we feel calm. Music or sound



G. Mazzola et al., *Making Musical Time*, Computational Music Science,
https://doi.org/10.1007/978-3-030-85629-8_6

process to our ear makes us feel a certain way. We will discuss rhythmical genealogy, from the biological perspective, and in particular from the neurological aspect, namely how the sound is processed to our ear and brain, and we explore how to understand the functions of local and global meters in the construction of time.

To this end, we first introduce the theory of Ravignani in his paper *The Evolutionary Origins of Rhythm: A Top-down/Bottom-up Approach to Temporal Patterning in Music and Language.* The top-down and bottom-up concept is frequently used in linguistics and borrows from psychology. A bottom-up process is that when we hear someone talking, we understand the small units (words), then the word combination becomes phrases, ending with the whole concept. Translated to music, local meters 'grow' into global meters.

On the other hand, top-down comprehension means to understand a sentence in a context based on previous knowledge. In music, it could be understood as feeling the groove/global meters (e.g. jazz as a style) before knowing the details/local meters (e.g. local swing/walking bass pattern).

To understand the difference between these two approaches, let's discuss concrete processing of sound[1].

Sound is created by differences in air pressure of waves traveling through air. However, our perception of sound is not directly related to differences in pressure. Our brain cannot receive this information directly. Instead, the human body uses a series of mechanisms to codify this information, provide the code to neuronal cells, and send this neural code to the brain [105].

Music is first translated into electrochemical information via the cochlea's hair cells in the ear. Once this information is received by the neurons, it is sent to the auditory cortex in the temporal lobe of the brain. The auditory cortex translates this information into what we perceive as pitch. After this point, it is unclear exactly how the brain is able to translate pitch, rhythm, articulation, and other aspects of music into a profound cognitive and emotional experience. However, it is clear that listening to music affects many parts of the brain, influencing parts of the brain responsible for pleasure (i.e., the nucleus accumbens) and emotion (e.g., the amygdala and hippocampus) [103].

When this information is processed in the temporal lobe where the signals construct chemical awareness, however, from listening ...the representation of the whole body/brain, which is known as the homunculus, as connected to the auditory rhythm perception is not known. We don't know how we connect sound perception to the movement...Our cognitive machine generates meaningful signs and messages. The sound is transformed into a semiotic entity, and we don't know how this is accomplished.

Given an example of typical bottom-up strategy: Ferdinand de Saussure's syntagm/paradigm dichotomy takes a central role in providing a semiotics of music. This dichotomy uses the rhythmical organization of objects to create

[1] Regarding other biological rhythm generators such as breathing, heartbeat, walking, refer to Section 6.1.2.

meaning. It thereby explains why rhythm is an important factor for the life's sense. In particular, rituals are relying on this construction. This dichotomy is based on two axes of meaning: the syntagmatic and the paradigmatic. The syntagmatic axis describes the position and placement of symbols (particularly their placement in time) in a succession of events, while the paradigmatic axis concerns groups of signs with similar meanings (called associative fields). The syntagmatic axis usually deals with tempi, rhythms, and note order while the paradigmatic axis deals with the use of similar melodic or harmonic patterns to create or resolve tension. In his work, Roman Jakobson suggested that the projection of the pragmatic axis to the syntagmatic axis is what generates poetical meaning. He called this process of (communication, art, language) the poetic function. The poetic function of communicative symbols is based on intentionality, and focuses on the importance of choosing signs carefully [103]).

On the other hand, the top-down strategy is present from the conductor's perspective. The conductor is given this creative license—to oversight the group to ensure that a common language is understood throughout the whole ensemble. The language of the group, however, may be semi-dependent on the conductor's interpretation of the music. That is, the conductor's speech, his or her individual patterns of musicality, will influence what the individual members of the ensemble must accept as a common language [103]. The conductor, in this way, is able to creatively express his or her musicality top-down through the ensemble. Despite this, each player still adheres to their own level of speech in that no other musician can play in the exact same way.

Soloists in particular are given an opportunity to express their musical speech, even to the extent of disregarding the conductor. Despite being most prominently observed in the conductor and soloists, all performers in the ensemble possess their own musical speech. It is the combination of these unique musical identities playing in concert with one another that gives rise to the incredible diversity of different, distinct ensembles and performances, even given the same orchestration. It is this aspect of the Saussurean speech/language dichotomy of music that makes it possible to produce creative expression of individuality through music. Understanding the language/speech dichotomy in music is crucial for the creation of a meaningful interpretation [103]. The top-down strategy is realized by the speech/language dichotomy.

In his research, Ravignani investigates the evolutionary and biological bases of time perception and rhythm cognition, and their role in the origins of music and speech. He argues that the idea that music and language both rely, among others, on the human cognitive ability for rhythm—intended as "patterning over time"—is unclear in the following areas: "when and how humans evolved rhythmic abilities, whether these were direct results of pressures for language or music, or only by-products of other evolutionary processes, and the degree of overlap between rhythm in language and music cognition." Also, rhythm is a cross-disciplinary concept, broadly employed not only in musicology and linguistics, but also in cognitive psychology, biology, and physics. In his talk, Ravignani presents his ongoing work, trying to unveil the evolutionary bases of time patterning, and to "build a unifying, theoretical framework for rhythm, transcending specific disciplines." He further argues how a broadly comparative approach, comparing different animal species, including humans, can inform us on the evolutionary origins of the cognitive ability to process rhythm [132]. He describes parallel and complementary lines of research to understand the origins of rhythm. First, he introduces a theoretical framework providing a common platform for the interdisciplinary study of rhythm in music and language [133]. The framework he proposes, the "hierarchy of coupled oscillations," attempts to connect human music, dance, and phonology on one side, to occurrences of synchronization and chorusing in non-human animals, as in bugs, on the other side. Second, he presents some preliminary results on group synchronization in "an agent-based model of chorusing, aimed at investigating rhythmic abilities in pre-musical hominids." In particular, he suggests how rhythmic complexity can emerge from simple, local interactions. This bottom-up approach is particularly useful in finding out which rhythmic features can exist without centralized processing [133].

Ravignani's ongoing experiments with this "Prima-Drum" will serve to investigate the temporal nature of drumming patterns in chimpanzees, compare them to those found in music and language, and ascertain chimpanzees abilities in discriminating and copying human-produced rhythms, as described in phonology and music theory research relatives: the chimpanzees.

Original quote:

> They naturally 'drum' in the wild. Ravigani describes the development and pilot testing of "a new tool to sonify apes' movements" [132]. Ongoing experiments with this "Prima-Drum" will serve to: "(i) investigate the temporal nature of drumming patterns in chimpanzees, (ii) compare them to those found in music and language, and (iii) ascertain chimpanzees' abilities in discriminating and copying human-produced rhythms (as described in phonology and music theory research)".

More recent studies by Ravignani include the following new insights and hypotheses:

In a paper [134], he reports that a female harbor seal pup developed clear timing reactions to tests with rhythmical acoustic signals even though it had never learned that from its other.

Ravignani has also described mathematical models of animal communication of temporal events in [135], and found out that the period of turn-taking is around 229 ms. Such a period has also been discovered for chimpanzee lip-smacks [125]. The conclusion is that our turn-talking rhythm could stem from this animal type of movement. We don't follow this hypothesis as the coincidence might also have a low-level anatomical rationale or be an insignificant coincidence. Ravignani also stresses the interaction of neuronal oscillators in the rhythmical collaboration of groups of animals, such as fire flies who produce their lights in congruence, like a harmonious drum orchestra, the synchronous flashing of fireflies.

In a third paper [136], they investigated the connection between vocal perception and body movements in animals and could prove a strong connection between these two expressivities.

The fourth paper [137] is very interesting as it supports two theories about the genealogy of human rhythm. The first theory argues that the human fetus experiences rhythmical events of the mother, when she walks, or breathes or even hearing her heart beat. The second theory is even more fundamental. They argue that the human two-pedal walking is unique in its two-periodic regularity. No other animals can move in in such a regular manner. From this anatomical fact of human movement, the theory infers that dance and music could have been developed. This last conclusion also relates to the sociological aspect of rhythmical genealogy.

### 6.1.2 Sociological Genealogy

**Summary.** In this section, we discuss the social genealogy of rhythm, that added value to the biological basics. We refer to recent cognitive and psychological experiments regarding the social germs in human rhythmical expressivity.

– $\Sigma$ –

Theorists of rhythm, e.g., in the interesting book *The Philosophy of Rhythm* [23], agree that there are biological oscillators (heart beat etc.) that may provide us with an original sense of rhythm, but as they argue rightly, animals also have those oscillators without developing a sense of rhythm. They observe [23, Ch. 12] that "in fireflies, frogs, crabs, and crickets, interpersonal entrainment is not a flexible process which can be deployed consciously in order to meet a specific goal." Therefore humans seem to be equipped with an additional rhythm 'generator'. However, it is not clear whether humans have an additional *biological* generator. The alternative is that the human sociological environment might be this additional generator. We do not decide here in how far this sociological germ would also be of biological nature, let us simply focus on the phenomenology.

In 2013, experiments regarding the social germs of rhythm have been reported by Daniel Bowling, Christian Herbst and Tecumseh Fitch [16]. This is their result:

> Using a paradigm in which participants read aloud with and without partners, we show that synchrony results in greater regularity of durational intervals between words, and that males and females are able to synchronize their temporal speech patterns with equal skill. These results demonstrate that the shared goal of synchrony can lead to the development of temporal regularity in human vocalizations, providing evidence for the possible origins of musical rhythm in social interaction.(...)This suggests that increased regularity of intervals between words arose specifically to facilitate synchronization, presumably by

allowing participants to accurately predict the timing of their partner's speech and coordinate their behavior accordingly.

In these experiments, temporal regularity, which is the elementary component of rhythm (see Section 6.2), results from a social vocalizing interaction. It seems that from this behavior an increased efficiency and understanding is produced when humans interact in temporal event sequences. Let us recall that poetologist Roman Jakobson has also stressed that the syntactically (and correspondingly rhythmically) regular distribution of associative fields in poetry is a generator of what he coined "poetical function", see [103, Ch. 9.1.4]. This gives the rhythmical component in human sociology an additional semantic signification: it helps create meaningful signs in time.

**Impact.** Rhythm seems to have its biological and social genealogy, whose combination generates the human reality of rhythm. Its gestural component being manifest, it is also remarkable, especially in the social rationale it embodies, that rhythm is not only a formalism, but also generates semantic layers of music on an elementary level, probably one of the core values of social phenomena: rhythm means something, and this might be characteristic for the exclusively human capability of creating rhythms with an aim, with the intention to signify. Here, the rhythmical construction of time generates an existential semantics, a level of temporal reality that transcends pure form.

– $\Theta$ –

## 6.2 Local Meters

**Summary.** Local meters are introduced following ideas of Hugo Riemann and from Ray Jackendoff's and Fred Lerdahl's Generative Theory of Tonal Music as elementary structures of onsets. The time domain is a prototype of a highly ambiguous dimension where intrinsic structures, such as notes, are mixed with extrinsic structures, such as bar-lines. The former is part of the musical material, the latter is part of the abstract time framework. We discuss the difference by exploring the example of Jackendoff's and Lerdahl's approach in [94, Vol. I].

– $\Sigma$ –

The reader should, however, be aware that we cannot include a philologically valid analysis of these works here; our scope is a modest one: to present the thread between these historically important outlets and our own approach as it really happened. The interest of such a tie is to start reflecting on the incredibly fuzzy state of conceptualization in music.

### 6.2.1 Riemann's Weights

**Summary.** Hugo Riemann's approach in [139] is discussed, stressing his conceptual specifications of meter and rhythm, in particular the weight idea.

– $\Sigma$ –

Riemann's concept of rhythm deals with a regular division of time ([139, p.2]):

> *Freilich ist aber doch wohl kein Zweifel, daß in der* erkennbaren Gleichmäßigkeit *der Zeitteilung wirklich ein besonderes lustgebendes Moment liegt und zwar eben gerade dasjenige, welches man Rhythmus nennt...*

The temporal period defining a rhythm is not an abstract one, it is related to the concrete music events of the given composition ([139, p.8]):

> *Welches in jedem Einzelfalle das Grundmaß [des Rhythmus, G.M.] ist, ergibt sich wie bereits betont, nicht aus dem abstrakten absoluten Mittelmaß, sondern aus dem der konkreten Melodie selbst.* Die Zählzeiten *(Schlagzeiten, rhythmischen Grundzeiten)* gewinnen unter allen Umständen erst reale Existenz durch ihre Inhalte.

For Riemann, rhythm is a result of grouping musical events to equitemporally distributed units, and to relate the grouped events to the units as relative instances (*Relativität der rhythmischen Qualität* [139, p.7]). This is why we have chosen the definition of a rhythmic weights as described in section 6.3. This is also stressed in the explication of rhythmical unity ([139, p.9]):

> *... als Inhalt einer den Pulsschlag des Rhythmus bildenden Einheit ist prinzipiell eine Mehrheit von wahrnehmbaren Erscheinungen* (also für die Musik: Tongebungen) *anzunehmen.*

However, a detailed study of Riemann's theory is not our scope and should be taken up in the framework of mathematical music theory as it has been taken up in motif theory with Rudolph Reti's approach by Chantal Buteau [17].

In contrast to rhythmical grouping, Riemann views metrical quality as a differentiation among onsets (of notes) which comes from their roles and relations. This quality is stated as a weight ([139, p. 8]):

> *... wir operieren mit einem zweiten Grundbegriffe, dessen Feststellung wir uns nun zuzuwenden haben, demjenigen des* verschiedenen Gewichtes der Zeiten, *der* **metrischen Qualität**.

The weight is meant to be a metaphor for the degree of importance:

> *... die Unterscheidung der einander folgenden Zeiteinheiten in wichtige und minder wichtige oder wie man zu sagen pflegt:* schwere *und* leichte ...

### 6.2.2 Jackendoff-Lerdahl: Intrinsic Versus Extrinsic Time Structures

**Summary.** The time domain is a prototype of a highly ambiguous dimension where intrinsic structures, such as notes, are mixed with extrinsic structures, such as bar-lines. The former are part of the musical material, the latter are part of the abstract time framework. We discuss the difference by the example of Ray Jackendoff's and Fred Lerdahl's approach in [61].

– $\Sigma$ –

In this section, we shall compare the metrical analysis of the Jackendoff-Lerdahl theory, Generative Theory of Tonal Music (GTTM), with what is proposed by the present mathematical music theory and its implementations in the Rubato platform's MetroRubette for metrical analysis. For a more generic discussion of the GTTM method, see [94, Ch. 21].

In order to perform a metrical/rhythmical analysis or interpretation of a musical composition, we must start by questioning the set of assumptions at the core of this form of composition. This basis is tied to the "neutral" work data on one hand, and to the analytical approach on the other. Every investigation of the neutral data presupposes a choice of relevant aspects (explain why). This choice occurs very early in the propaedeutic evaluation.

For example, in applying Rubette to the base assumptions in the first Viennese school, bar-lines, period limits, notes, and pauses are recognized. In a metrical/rhythmical analysis, the question of the bar-lines' relevance would be set forth: Will this type of event strongly influence our analysis? As is shown in the MetroRubette concept, the answer is not necessarily "Yes" or "No", a fuzzy decision between these two extremal values is possible and reasonable.

Our fuzzy approach avoids a normative analysis, which would imply that we decide about the inclusion of bar-lines as external formal limits of metrical/rhythmic analysis. (Transition/elaborate/refder) This bed of Procrustes is what GTTM prescribes in its metrical structure analysis: Bar-lines are always instances of the metrical structure; this is implicit in the fourth rule 1 of metrical structure 2 (see Figure 6.1). Here are the four MWFRs:

- **MWFR 1** Every attack point (refer to Basic Music Technology the chapter explaining attack point in MIDI) must be associated with a beat at the smallest metrical level present at that point in the piece.
- **MWFR 2** Every beat at a given level must also be a beat at all smaller levels present at that point in the piece.
- **MWFR 3** At each metrical level, strong beats are spaced either two or three beats apart. (This rule is recognized as idiom-specific in GTTM.)
- **MWFR 4** The tactus and immediately larger metrical levels must consist of beats equally spaced through the piece. At subtactus levels, weak beats must be equally spaced between strong beats.

Worse than that: even non-existent onsets are inserted in determining levels in order to enforce periodicity where the text does not show such. This is a consequence of (for example) the axiom MWFR3 which asks that successive metrical levels have duration (period) ratios 1:2 or 1:3. All other ratios are forbidden! The analysis of eight bars from Mozart's Jupiter Symphony is a typical example, see Figure 6.1.

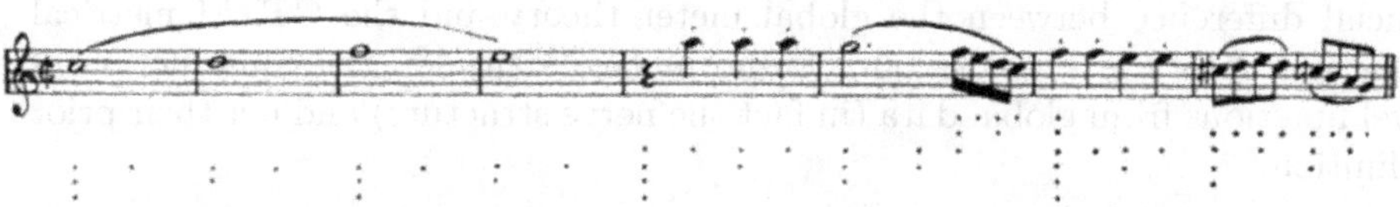

Fig. 6.1: The metrical hierarchy in the finale's eight-bar theme of Mozart's Jupiter Symphony, according to the GTTM, [94]. Reprinted with permission of The MIT Press.

It shows that the whole notes in the first four bars have to be enriched by half-note points in order to cope with the succeeding onset structure. It also shows that the bar-lines are points even when there is no such note-onset. Such a procedure is not only rigid, but it is also dangerous since the complexity of meter/rhythm is destroyed by secondary instances such as bar-lines, and by a spurious subdivision dogma (1:2, 1:3). And it enforces a norm which only cements bad popular taste instead of representing possible compositional intentions.

But even in tonal music, also in modern tonal music, rhythm and metrics are much less tied to bar-lines than the formal score notation suggests. For instance, in contemporary jazz, e.g., Steve Coleman's CD *The Sonic Language of Myth*, we recognize extremely complex poly- and micro-rhythms that escape bar-lines and would be eliminated by the said normative standardization as being non-metrical or non-rhythmic music.

Moreover, the GTTM casting of metrical structure makes evident a remarkable drawback of non-computerized musicology: The effective usage of the following metrical analysis method (21.2) would have been fictitious without computers because of its combinatorial complexity. We in fact argue that sometimes, a normative approach is only propagated because operational control of analysis would break down without such a "terrible simplification." So the simple and wrong is preferred to reality. Which would possibly force one to wait until the power of tools has reached an adequate level. For example (refer to *Take Five* as 'destroyed' rhythmically as a contradict example–impossible division of five) During the design period of the workstation Rubato it was therefore mandatory to ask for the following condition:

*We need concrete tools which are adequate to the analyzed work, and which yield an immanent analysis, not merely abstract or brutally normative principles.*

Despite our critical review of the GTTM method [100, Ch 16.1], the idea to cover the onsets of a given composition by "levels" has been a fruitful input for our global metrical perspective. It gave us the motivation to introduce global meters from the concept of a local meter, which were introduced in [94], and from the insight that simultaneous presence of several local meters in a piece of music is the rule and not the exception (see Figure 13.12 in [94]). The crucial difference between the global meter theory and the GTTM metrical structure theory is—besides our non-normative attitude—the construction of level functions from global data (in fact the nerve structure) and not their prior definition.

### 6.2.3 The Formal Theory of Local Meters

A rigorous and more general definition of local meters is given in [93, Section 7.2.2]. Here we give a rigorous, but somewhat simplified definition, which is nevertheless a representative picture of the local conceptualization of time. All theorems that will be presented here without proof are proved in [93].

**Definition 1** *A local meter $l$ is a (usually finite) subset $m \subset Onset$ of the space of onsets Onset with real number coordinates, having the form $m = \{o_0, o_1 = o_0 + p, l_2 = l_1 + p, \ldots o_l = o_0 + lp\}$, a sequence of $m+1$ onsets that are generated by a successive addition of a positive* period $p = p(m)$, *ending with the $l$-fold addition, $l$ being called the* length $l(m)$ *of $m$. Length $l(m) = 0$ means that we have a single onset. Except for this case, length and period are uniquely determined by $m$.*

Local meters are the most elementary or atomic structures of the metrical and rhythmic structure of time. Their justification is also given from our preliminary discussion of biological and social rhythmical genealogy as they formally describe the simple repetition of an onset with a constant period.

A slightly more general concept is what we call a *local rhythm.* This consists of three components: first, a local composition $G$, i.e., a set of sound events which all share the onset coordinate, but may have many others, such as pitch, loudness, duration, voice, etc. We call it the rhythm's *germ.* Second and third, we are given a period $p$ in onset time, as well as a non-negative integer length $l$. From these three ingredients we generate the union

$$R(G, p, l) = \bigcup_{i=0,\ldots l} T^{ip} G,$$

i.e., the union of all translates of $G$ by multiples of the period onset. All other parameters, except onset, are not affected. If the germ is just an initial onset, we get the local meter concept back as a special case.

**Impact.** Starting from historical roots of precise definitions of (local) meters, we presented a contemporary mathematically valid definition of these concepts that was described in [93, Part III/7] after having been implemented in the RUBATO® software in 1992.

– $\Theta$ –

## 6.3 Global Meters

**Summary.** Global Meters are manifolds of local meters and define nerves of such coverings and their associated homological concepts. The hierarchic structure of global meters gives rise to topologies and associated weights, i.e., numerical functions that contain information about the local connection of onsets within their topological neighborhoods. Metrical topologies are the theoretical background for the MetroRubette module for metric and rhythmic analysis in the Rubato software discussed in Basic Music Technology [105].

– $\Sigma$ –

It is interesting that the intersection of two local meters does not generate a new structure:

**Proposition 1** *The intersection* $m \cap n$ *of two local meters* $m, n$ *is a local meter if it is non-empty.*

This means that the onset sets that are unions of local meters define a topology on the set of onsets. Of course, this is also true for local rhythms, namely looking at the local meter of the intersctetion onsets, together with the intersection of the two germs.

In particular, if we consider, for a given composition $K$ of onsets (as derived from a concrete composition by projecting its notes to the onset axis), the set-theoretically maximal local meters within $K$ generate a topology, whose minimal open sets are finite intersections of maximal local meters. This topology is called the *maximal local meter topology* on $K$. In particular, for any $x \in K$, one may consider its *simplex* $Sp(x)$, namely the set of all maximal local meters containing $x$. With this concept we have the following fact. It relates the simplices of points $x$ to their topological closures $\overline{x}$. We write $x < y$ iff $\overline{x} \subset \overline{y}$.

**Proposition 2** *For any two points* $x, y$ *in the composition* $K$ *of onsets, we have* $x < y$ *iff* $Sp(x) \subset Sp(y)$.

This means that the topological 'dominance' of onsets is given by the size of their simplices. This allows for a quantitative measure of the onsets' relevance in a given composition $K$. We define the rhythmical weight w(x) of an onset $x$ by the formula

$$w(x) = \sum_{m \in Sp(x), l(m) \geq q} l(m)^p$$

for a given *profile* number $p$, and a minimal length $q$ of admitted maximal local meters. The limit $q$ tells us that below the length $q$, maximal local meters are not considered, and that the growth of local meter lengths is taken into account by the profile's exponent $p$.

This quantitative evaluation of global metrical structures of a composition have been implemented in the Rubato's component MetroRubette. Figure 6.2

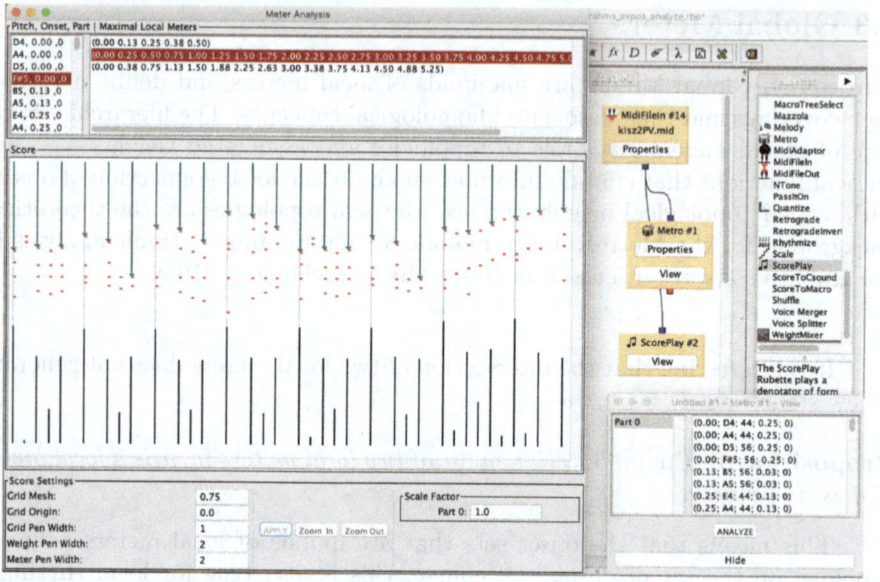


Fig. 6.2: A screenshot of Rubato's MetroRubette weight from Schumann's Kinderszene No. 2.

shows the weight graph (vertical black lines from the bottom) and one of the relevant maximal local meters (green, from the top) in Schumann's Kinderszene No. 2, Kuriose Geschichte, whose notes are represented by red points.

## 6.4 Metrical Topologies

**Summary.** Metrical topologies are topological spaces associated with Grothendieck's construction of the functor of sober spaces.

– $\Sigma$ –

The above concept of the simplex $Sp(x)$ of a point $x \in K$ is an interesting concept of topology. In fact, one may define a space $X^s$ for every topological space $X$, namely the associated *sober* space. Its elements are the closures of the points $x \in X$. Its open sets are the sets of closures of points that intersect with original open sets. It is interesting to learn that in our case, the projection $X \to X^s$ sending $x \in X$ to its closure is in fact the quotient topology, i.e., the open sets in $X^s$ are those sets whose inverse image is open in $X$. In other words, the map $x \mapsto Sp(x)$ is the map to the sober space, and the dominance of points translates to the subset relation of simplices.

We mention these facts to show that the weights of onsets as defined in the previous section are intrinsically related to the sober topology associated with the maximal meter topology of a composition $K$ of onsets.

## 6.5 Rhythms

**Summary.** This generalizes the metrical conceptualization to include more musical parameters beyond onset.

– $\Sigma$ –

The local rhythms as defined above are an enrichment of the metrical topologies. For a given musical composition, its projection onto its onsets generates the maximal meter topology, and via inverse images, we may consider rhythms that are associated with maximal local meters. In other words, maximal local meter topologies are projections of rhythmical topologies.

This construction of musical time structures starting with local meters is a strong technical tool that allows for a quantitative and qualitative investigation of phenomena in compositions that are generated by simple local meters. We should add that the metrical and rhythmical analysis of musical compositions by means of this theory has revealed important time architectures, which were not recognized in any precise way before, in famous European compositions by Schumann and Chopin, see [95, Chapters 43.3 and 45]. More recently, the rhythmical analysis of Johannes Brahms' first piano sonata revealed sensational new insighs into Brahms' out-of-meter approach already in his early works [104, Chapter 6.4].

This approach to musical time originates in the gestural roots of rhythmical concepts, as presented in Section 6.1. But it evolves into a sophisticated conceptualization of global metrical and rhythmical phenomena, converging to a topological picture that is also quantified for music software and thereby reveals deep hidden rhythmical constructions by famous composers of European master works.

It is nevertheless still problematic to disconnect this mathematical theory from its gestural roots. The efforts of performers should address the reembodiment of such topological constructions in their gestural performative utterance.

**Impact.** We developed the topological architecture of global rhythmical structures, an approach to musical time that originates in the gestural roots of rhythmical concepts.

– $\Theta$ –

# 7

# Structures of Organized Time

**Summary.** Harald Krebs', Guerino Mazzola's, Gérard Assagay's, and Jason Yust's contributions to hierarchical time concepts.

– $\Sigma$ –

## 7.1 Harald Krebs

**Summary.** Harald Krebs introduces an interesting conceptualization of the simultaneous presence of different metrical structures, which he calls metrical consonance and dissonance. We discuss his approach and interpret it as an anticipation of our more recent topological theory of local and global meters.

– $\Sigma$ –

Krebs writes in Chapter 2 of [73]:

> I define the meter of a work as the union of all layers of motion (i.e., series of regularly recurring pulses) active within it. The layers that contribute to the meter of a work can be divided into three classes: the pulse layer, micropulses, and interpretive layers. The pulse layer is the most quickly moving pervasive series of pulses, generally arising from a more or less constant series of attacks on the musical surface. (The omission of a few pulses here and there does not seriously disrupt the pulse layer once it is clearly established.) More quickly moving layers, or "micropulses," may intermittently be woven into the metrical tapestry of a work as coloristic embellishments. Of greater significance are series of regularly recurring pulses that move more slowly than the pulse layer. These allow the listener to "interpret" the raw data of the pulse layer by organizing its pulses into larger units. The pulses of each "interpretive layer" subsume a constant number of pulse-layer attacks; an interpretive layer can therefore be characterized by an integer denoting this constant


G. Mazzola et al., *Making Musical Time*, Computational Music Science,
https://doi.org/10.1007/978-3-030-85629-8_7

quantity. I refer to this integer n as the "cardinality" of the layer, and to an interpretive layer of cardinality n as an "n-layer."

I conceive of interpretive layers not as components of an abstract, "given" metrical grid, but as perceptible phenomena arising from the regular recurrence of musical events of various kinds. Interpretive layers often arise from a regularly spaced succession of what Lerdahl and Jackendoff call phenomenal accents—"event[s] at the musical surface that give emphasis or stress to a moment in the musical flow." Dynamic accent, shown in scores by markings like *of, fz, rf,* and by various hairpins and carets, is a particularly obvious type of phenomenal accent. Such accentuation is, however, also produced by the placement of long durations among short ones (agogic or durational accents), thickly textured events among more thinly textured ones (density accents), by dissonant events among consonant ones, by registral high and low points (registral accents), by the affixing of an ornament to a note, by changes in harmony and melody (new-event accents)—in short, by any perceptible deviation from an established pattern. A succession of accents occurring at regular intervals—that is, the high-lighting of every nth member of the pulse layer—results in the establishment of an interpretive layer of cardinality n.

(...)

*Interaction of layers*: In the above examples, as in most tonal music, more than one clearly perceptible interpretive layer is active at once. These layers interact in one of two ways: they align (or "nest," to use Joel Lester's term) or they do not. A set of layers aligns when each pulse of each interpretive layer coincides with a pulse of every faster-moving layer. This state of alignment always exists when only one interpretive layer is imposed on a pulse layer. Two or more interpretive layers, however, may or may not align with each other. I refer to the metrical state resulting from the alignment of interpretive layers as "metrical consonance." The appropriateness of this metaphorical use of a term traditionally applied to pitch rests on the literal meaning of the word "consonance"; metrical consonance exists when pulses sound together. The metrical state arising when interpretive layers do not sound together can, applying the same metaphor, be termed "metrical dissonance."

Before we investigate these two states more closely, let us briefly consider the adjective "metrical." Its aptness is obvious when states of alignment or non-alignment result from the interaction of layers whose pulses are metrical beats, or beats above the level of the bar line (hypermetrical beats, or hyperbeats). But what of sets of layers whose pulses are submetrical, in some cases not even notationally present (for instance, the micropulse arising from superpositions of eighth notes and triplet eighth notes)? It might be argued that sets containing such layers should not be termed "metrical" consonances or dissonances. Since,

> however, I am defining meter as the union of all layers of motion active within a composition, I include even these submetrical interactions under the terms "metrical consonance or dissonance."
> (...)
> Metrical *consonance*, the normal state of pre-twentieth-century tonal music, involves the aligned or nested presentation of interpretive layers whose cardinalities are multiples/factors of each other.
> (...)
> As nineteenth-century theorists already hinted, there are *two basic ways in which nonalignment or dissonance between regular layers of motion can occur:*
> - First, dissonance can be formed by the association of at least two interpretive layers whose cardinalities are different and are not multiples/factors of each other.
> - The second basic type of dissonance involves the association of layers of equivalent cardinality in a nonaligned manner; that is, dissonance of this type does not depend on the association of noncongruent layers, but merely on the different positioning of congruent layers.

Krebs here presents a conceptualization of metrical phenomena on different levels. Clearly, however, he is talking about what we introduced in Chapter 6 as local meters, he describes them as: "series of regularly recurring pulses". He distinguishes three categories of such meters: pulses, micropulses, and interpretive layers. This classification is enabled in our own approach as soon as one differentiates periods and lengths of local meters, and also the musical events that one considers (bar lines, pauses, notes of different durations, loudness, etc.).

The more interesting feature in Krebs' theory is the concept of metrical consonance or dissonance. He writes about layers in alignment: "A set of layers aligns when each pulse of each interpretive layer coincides with a pulse of every faster-moving layer." This is what he conceives as consonance. This can also be interpreted as the amount of common events among a number of metrical layers. If they align, their intersection is large, else, it is small or even empty. This understanding is congruent with what we have discussed under the title of metrical topology. Disjoint open metrical sets would be in dissonance, while the heavy intersection would be a phenomenon of consonance. Of course, our topological approach, including all the numerical criteria deduced from simplices of onsets and their associated weights is much more precise and explicit than the consonance/dissonance dichotomy of Krebs. But his approach seems very interesting, even as a kind of anticipation of our topological perspective.

**Impact.** Harald Krebs proposes a conceptual approach that looks in fact like an anticipation (despite its mathematically imperfect terminology) of our topological theory of rhythm as described in Section 6.4.

– $\Theta$ –

## 7.2 Guerino Mazzola

**Summary.** Guerino Mazzola developed and implemented a hierarchical formalism for tempo. We give a summary of its architecture and relation to the concept of a musical time.

– Σ –

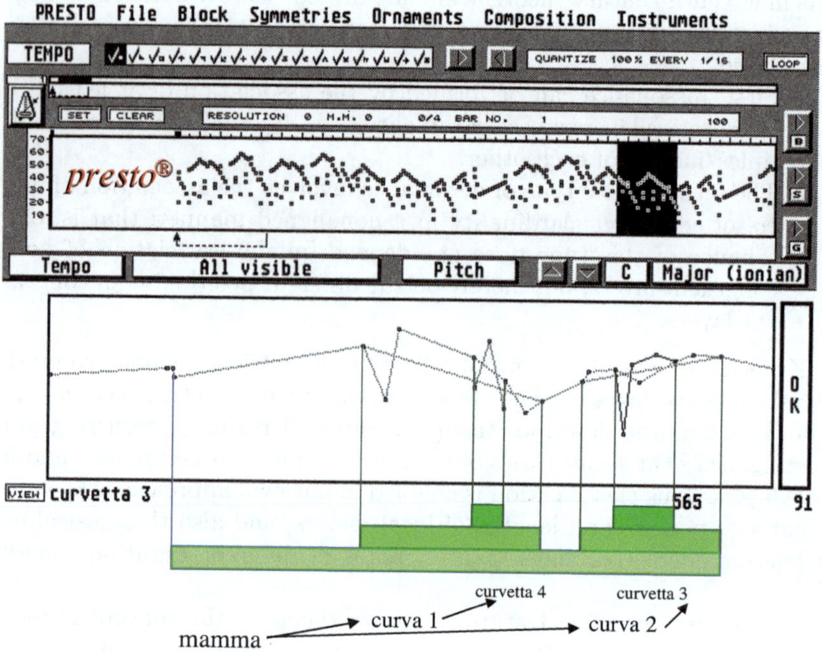


Fig. 7.1: The tempo hierarchy of an orchestra is shown. The mother tempo curve (mamma) is ramified into two daughters (curva1, curva2) which in turn have one daughter each (curvetta4, curvetta3, respectively). This interdependence of local and global tempi is used to differentiate roles in orchestral time control.

The construction of musical time involves among others the construction of the transformation of symbolic time to physical time, which is defined by tempo. If this transformation is described by the map

$$P_T : SymbolicTime \to PhysicalTime : E \mapsto e(E),$$

the tempo at a symbolic time $E$ is defined as the inverse $T(E) = \frac{1}{de/dE}$. Traditionally, this defines the tempo curve $T(E)$. However, this is not what happens in realistic tempo constructions. In reality, tempo curves are built following a hierarchical deployment of tempi in a stemmatic genealogical setup. The underlying principle was developed in performance theory, as described in [93, Ch. 38]. If we are given a tempo curve $T(E)$, the curve between two symbolic onsets $E_1 < E_2$ might be refined by a "daughter tempo curve" $T^*(E)$ with the property that the physical time between $e(E_1)$ and $e(E_2)$ under $T^*$ would coincide with the physical time between theses two symbolic times under $T$. This means that

$$\int_{E_1}^{E_2} 1/T(E)dE = \int_{E_1}^{E_2} 1/T^*(E)dE.$$

We obtain a new tempo curve $T'(E)$ that coincides with $T(E)$ outside $[E_1, E2]$ and with $T^*(E)$ inside this time interval. This refinement procedure may be repeated for other time intervals, but also adding more refinements to the curve $T'(E)$. Figure 7.1 shows such a hierarchy for an orchestra, as it was implemented in the composition software *presto*®. This means that the musical time as built from tempo is a hierarchy of local tempo curves.

**Impact.** Mazzola's theory of tempo hierarchies, as implemented in his *presto*® software, describes tempo as a multidimensional architecture of local time structures, an approach that contradicts the linear time principle.

– $\Theta$ –

## 7.3 Gérard Assayag

**Summary.** Assayag in his OpenMusic software applies a hierarchical concept of musical time. We look at its formal shape and the compare it to Jason Yust's time hierarchies.

– $\Sigma$ –

Rhythmic Trees (*RT*s) are the base of the rhythmic representation in Assayag's music software OpenMusic. They must be understood as an alternative description of symbolic rhythmic structures using traditional music notation, and an external specification for container objects.

- **Syntax** An *RT* is defined as a pair $(D\ S)$ where $D$ is an integer ratio ($> 0$) and $S$ is a list of $n$ elements. Each element in $S$ can be either an integer or a *RT*. Here is an example corresponding to this syntax:
  (2 ((1 (1 1 1 1)) (1 (1 1 1 1)))).
  We have here a nice example of a doubly circular form[1]. To begin with, given a form $F$, a list form over $F$ is the circular form

$$List(F) \underset{Id}{\longrightarrow} \mathbf{Colimit}(Item(F), Terminal)$$

[1] Forms and denotators are basic concepts of Mazzola's concept architecture for mathematical music theory [94].

with the terminal form $Terminal \xrightarrow[Id]{} \mathbf{Simple}(\mathbb{Z})$ in order to formalize the end of a finite list and to indicate its length with the integer of the coordinator module $\mathbb{Z}$. The form $Item(F)$ is a Cartesian product

$$Item(F) \xrightarrow[Id]{} \mathbf{Limit}(F, List(F)).$$

Building on this form, we define the form $RT$ by

$$RT \xrightarrow[Id]{} \mathbf{Limit}(DurationValue, List(RT))$$

and a simple duration form

$$DurationValue \xrightarrow[Id]{} \mathbf{Simple}(\mathbb{Q}).$$

- **Semantics** For a given $RT = (D, S)$, $D$ expresses a duration and $S$ defines a group of proportions of $D$. For instance, by taking as unity the quarter note we have for $RT = (1, (1, 1, 1, 1))$ the rhythm shown in Figure 7.2.

Fig. 7.2: Rhythm for $RT = (1, (1, 1, 1, 1))$.

$RT$s allow us to represent, in a homogenous way, different types of musical objects. Polyphonies, voices, measures groups, etc. are expressed as $RT$s. When the value $D$ is at the measure level, we express it in whole note units. For example, the $RT$ for the next rhythm will be $(3/4, (2, 1))$, see Figure 7.3.

Fig. 7.3: $RT = (3/4, (2, 1))$.

As it was defined, $S$ represents a sequence of proportions of $D$. The example in Figure 7.4 shows the case for $RT = (4/4, (1, 2, 1))$.

Fig. 7.4: $RT = (4/4, (1, 2, 1))$.

Until now, we have presented $RT$s where $S$ was a list of integer elements, representing notes or beats. For an $RT = (4/4, (1, (2, (1, 1, 1)), (1, (1, 1, 1))))$,

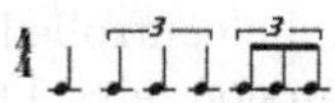


Fig. 7.5: $RT = (4/4, (1, (2, (1, 1, 1)), (1, (1, 1, 1))))$.

more complex elements appear as shown in Figure 7.5. The *RT*s contained in $S$ represent what we call *(rhythmical) groups.* In general, groups are graphically represented as beamed notes. As well as measures, groups can contain either notes or other groups. Here is an example (Figure 7.6) for the

$$RT = (4/4, (1, (1, (1, 1, (1, (1, 1, 1)), 1, 1)), 2)) :$$

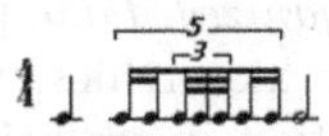


Fig. 7.6: $RT = (4/4, (1, (1, (1, 1, (1, (1, 1, 1)), 1, 1))2))$.

We extend the *RT* syntax in order to integrate rests and ties which will be represented respectively by negative numbers and floats. For

$$RT = (4/4, ((1, (1, 1)), (2, (1.0, 1, -2, 1)), (1, (1.0, -1, 1))))$$

we obtain the following rhythm (Figure 7.7):

Fig. 7.7: $RT = (4/4, ((1, (1, 1)), (2, (1.0, 1, -2, 1)), (1, (1.0, -1, 1))))$.

*RT*s may encode the metric intention in a rhythm. The two rhythms in the next Figure 7.8 are encoded by different *RT*s $RT1 = (2/4, (2, 2, 2, 2))$ and $RT2 = (2/4, ((1, (1, 1)), (1, (1, 1)))$.

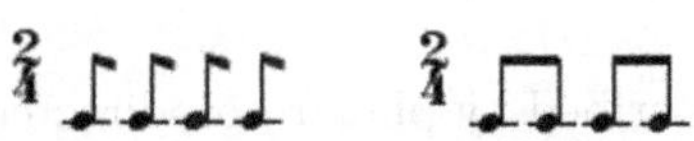

Fig. 7.8: *RT*s $RT1 = (2/4, (2, 2, 2, 2))$ and $RT2 = (2/4, ((1, (1, 1)), (1, (1, 1)))$.

The symbolic representation is essentially based on the hierarchical musical content. This allows an overall readability of the hierarchical structure and a symbolic format, which can be controlled by algorithms and other transformations such as inversion, recursion, etc. This is made possible by the fact that the format is the abstraction of the object itself. *RT*s are automatically encoded into optimal container structures.

## 7.4 Jason Yust

**Summary.** Jason Yust has developed a hierarchical formalism for organized time. We give a summary of its architecture and relation to the concept of a musical time.

– $\Sigma$ –

In his fascinating book *Organized Time* [171], Yust describes an organizational principle that generates hierarchies, which can be applied to tonal, formal, and rhythmical organization in traditional Western musical composition. The organization follows a graph-theoretical display, which we show in Figure 7.9, an example from Yust's book on page 14.

Fig. 7.9: an example from Yust's book on page 14. The metrical hierarchy of the theme from Haydn's *London* symphony no. 104. (a) with empty timepoints, (b) with the existing timepoints of notes.

This hierarchy is described by planar non-directed simple[2] graphs, where the vertices represent timepoints (real of imaginary), while the edges stand for time intervals, however without making their duration precise. These graphs are ideally maximal outerplanar (MOP in Yust's writing), i.e., they are planar,

[2] No loops, no multiple edges.

the have a Hamilton cycle[3] that is defining the outerplanar domain, and they are maximal with these properties.

Such a graph may be defined by a denotator form as follows: We first introduce the edge form, together with edge duration (which is absent in Yust's approach):

$$Edge : .\mathbf{Simple}(\mathbb{Q})$$

We then introduce a list form of triangular structures:

$$MOPList : .\mathbf{List}(GenTriangle)$$

The members of such a list are a colimit form of generative triangle structures:

$$GenTriangle : .\mathbf{Colimit}(Edge, Triangle)$$

i.e., either edges or triangles, where the *Triangle* form

$$Triangle : .\mathbf{Limit}(Edge, GenTriangle, GenTriangle)$$

is the limit of an *Edge* form plus its two generative triangles. This circular recursive construction terminates, when we are only given edges in the *GenTriangle* form.

This setup can evidently also written down as a 2-tree, where the two-fold ramifications are the two triangles in the *Triangle* form. The reader might make the exercise to apply our denotator formalism to describe the Haydn example in Figure 7.9.

Yust's temporal hierarchies are not definitions of musical time, the take it as a given reality, and they don't focus on the durations as rational numbers. His formalism also applies to tonal and formal aspects, the metrical/rhythmical aspect is just one of several possible hierarchies. Its essence is the binary subdivision process, as made evident by the 2-tree representation. It is similar to the Jackendoff-Lerdahl approach, where types of subdivisions are strongly limited. For Yust, subdivisions into more than triangles are termed "holes", a very selective point of view. Why not introduce higher polygons as standard subdivisions beyond triangles?

Here is a comparative diagram of the forms as proposed by Assayag and Yust. Both are circular and have the same depth, see Figure 7.10:

Both approaches are hierarchical and circular. They both do not define the basic time dimension, the take it implicitely (Yust) or explicitely (Assayag) as given. Their power is the important step from linear time to a hierarchy of times.

*It is remarkable that musical time is multidimensional, there are mother and daughter times, a whole family. This situation is also encountered in the*

[3] A Hamilton cycle of a graph is a cycle that traverses each vertex only once, except the start and end.

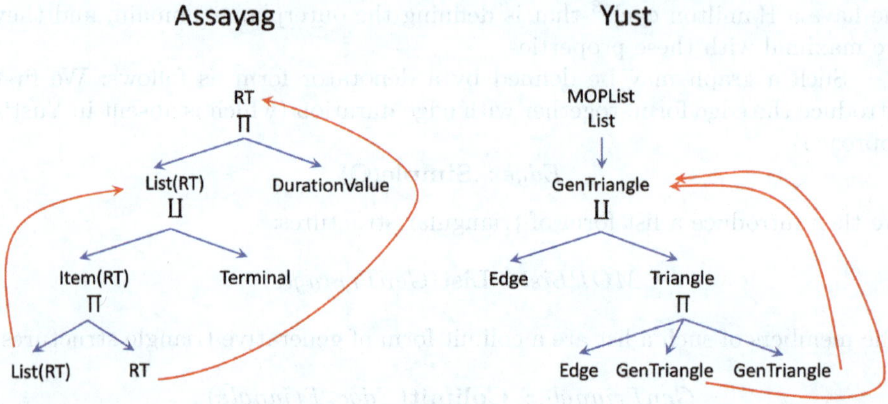


Fig. 7.10: The temporal forms as proposed by Assayag and Yust. Both are circular and have the same depth.

*stemma theory of musical performance, see* [95], *where the evolution of performative quality includes an evolution of tempo layers, as discussed in Section 7.2.*

**Impact.** Both, Assayag and Yust describe a circular hierarchical time architecture, Assayag also implements it in his Open Music software. Similar to Mazzola with his tempo hierarchies, they however do not deal with the definition of time as such, but stress the *constructive* multidimensionality of time in music.

– $\Theta$ –

# 8

# Musical Gestures

**Summary.** This theory was introduced by Guerino Mazzola and is described in detail in [96]. Here we give a summary of the theory that is relevant to the conceptualization of time.

$-\Sigma-$

In this chapter we shall present a concise, but somehow simplified version of the mathematical theory of musical gestures. We shall restrict our discourse to gestures in topological spaces, not the more general topological categories.

There are three components of this theory, which are most important for the present discussion: the definition of gestures, the construction of hypergestures with the Escher theorem, and the definition of gestural morphisms, which entails the concept of the category of gestures that is crucial for our temporal conceptualization in Chapter 11.

## 8.1 Gestures and Hypergestures

**Summary.** In Mazzola's musical gesture theory, a gesture is a representation of a directed graph (the gesture's skeleton) as a system of continuous curves in a topological space (the gesture's body). This enables the construction of topological spaces whose points are gestures, and therefore of hypergestures, i.e., gestures whose curves are curves of gestures.

$-\Sigma-$

Let us first review the category *Digraph* of directed graphs, short: digraphs. This is the basic category for algebra as well as for topology. In the naive setup, its objects are functions $\Gamma : A \to V^2$ from a set $A = A_\Gamma$ of *arrows* to the Cartesian square $V^2 = V \times V$ of the set $V = V_\Gamma$ of *vertices*. The first projection $t = pr_1 \circ \Gamma$ is called the *tail* function, the second $t = pr_1 \circ \Gamma$ is called the *head* function of the digraph. For an arrow $a$, the vertices $t(a)$, $h(a)$ are called its head and tail, respectively, and denoted by $t(a) \xrightarrow{a} h(a)$ or also

 
G. Mazzola et al., *Making Musical Time*, Computational Music Science,
https://doi.org/10.1007/978-3-030-85629-8_8

$a : t(a) \to h(a)$. A morphism $f : \Gamma \to \Delta$ of digraphs is a couple $f = (u, v)$ of functions $u : A_\Gamma \to A_\Delta, v : V_\Gamma \to V_\Delta$ such that $v^2 \circ \Gamma = \Delta \circ v$.

A more creative definition of this category is to view it as the category of presheaves[1] $ht^{@}$ over the small category $ht = \bullet \overset{h}{\underset{t}{\rightrightarrows}} \bullet$ with two objects and just two parallel morphisms $h, t$ between these points. This interpretation[2] implies that *Digraph* is a topos. In particular, it has a a final object $1 = t \circlearrowleft T$, which is embedded by the true morphism $T : 1 \to \Omega$ into a subobject classifier

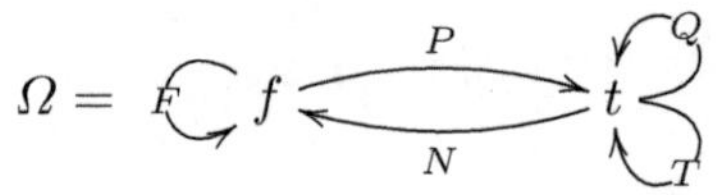


Every evaluation $\Gamma@\Omega$ describes the set of subdigraphs of $\Gamma$, together with its canonical Heyting logic.

In this chapter, we need a special subcategory of digraphs, the *spatial digraphs*. Such a digraph is associated with a topological space $X$ and denoted by $\overrightarrow{X}$. By definition, the arrow set is $A_{\overrightarrow{X}} = I@X$, the set of continuous curves $c : I = [0, 1] \to X$ in $X$, while the vertex set is $V_{\overrightarrow{X}} = X$, $h(c) = c(1)$, and $t(c) = c(0)$. A spatial morphism is a digraph morphism $\overrightarrow{f} : \overrightarrow{X} \to \overrightarrow{Y}$ canonically induced by a continuous map $f : X \to Y$. The subcategory of spatial digraphs and morphisms is denoted by *SpaceDigraph*. A spatial digraph is more than a digraph: it is also a topological digraph in the following sense. The set $A_{\overrightarrow{X}} = I@X$ of arrows of $\overrightarrow{X}$ is a topological space by the compact-open topology, and the head and tail maps $h, t : I@X \to X$ are continuous. Moreover, for a continuous map $f : X \to Y$, the arrow map $I@f : I@X \to I@Y$ is continuous.

Given a digraph $\Delta$ and a topological space $X$, a *$\Delta$-gesture in $X$* is a digraph morphism $\delta : \Delta \to \overrightarrow{X}$, i.e., a realization of the abstract vertices and arrows within a topological space, as shown in Figure 8.1. It is essential here to distinguish the curve parameter of a gesture from the time parameter, which intervenes in a number of common gestures. For example, when drawing a curve on a sheet of paper, this gesture $\delta : \Delta \to \overrightarrow{X}$ would have the arrow digraph $\Delta = \uparrow$ with $\uparrow = \bullet \to \bullet$ and the space $X = \mathbb{R}^2 \times \mathbb{R}$, whose first two coordinates denote the points of the paper surface, whereas the third denotes the physical time when a point at given parameter value is drawn on the paper sheet. We discuss the time parameter in the following example 1.

---

[1] For any category $\mathcal{C}$, $\mathcal{C}^{@}$ denotes the category of presheaves, i.e., contravariant set-valued functors, on $\mathcal{C}$. Our notation stems from the Yoneda embedding $y : \mathcal{C} \to \mathcal{C}^{@}$, which associates with an object $X$ in $\mathcal{C}$ the presheaf $@X$ which yields the set $@X(Y) = Y@X$ of morphisms $f : Y \to X$ in $\mathcal{C}$.

[2] Observe that $ht^{op} \xrightarrow{\sim} ht$, so it is also legitimate to view $ht^{@}$ as being built from the covariant functors on $ht$.

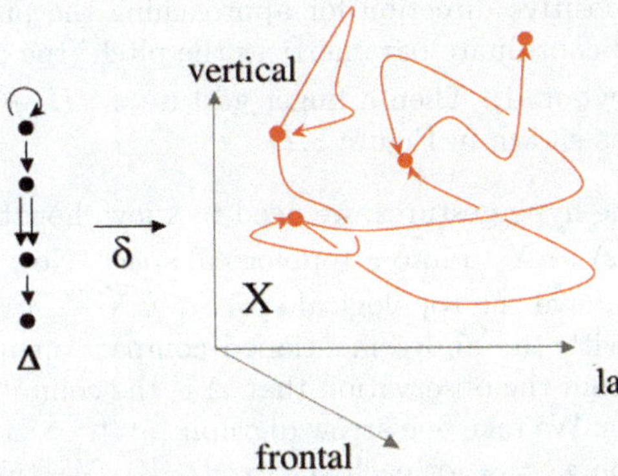


Fig. 8.1: A gesture in the ordinary 3-space.

Given two gestures $\delta : \Delta \to \overrightarrow{X}, \gamma : \Gamma \to \overrightarrow{Y}$, a *morphism f gestures* $f : \delta \to \gamma$ is a digraph morphism $f : \Delta \to \Gamma$ together with a spatial morphism $\overrightarrow{h} : \overrightarrow{X} \to \overrightarrow{Y}$ which commutes with $f$, i.e., $\overrightarrow{h} \circ \delta = \gamma \circ f$. This defines the category *Gesture* of gestures, and we have a projection $p : \mathit{Gesture} \to \mathit{Digraph}$, which sends the gesture $\delta : \Delta \to \overrightarrow{X}$ to the underlying digraph $\Delta$. So it is essentially a forgetful functor: it cancels out the spatial interpretation of the given digraph.

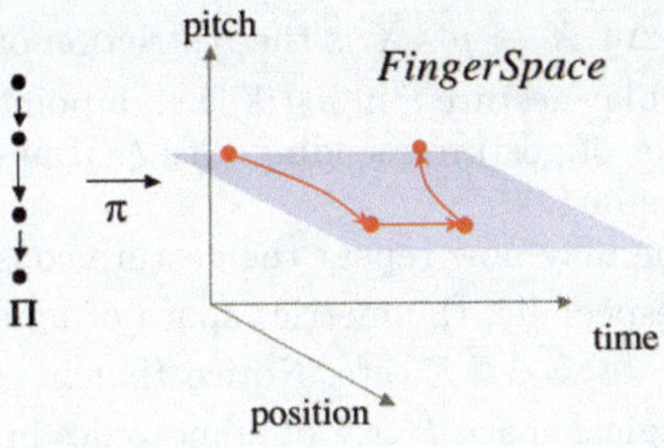


Fig. 8.2: An elementary finger gesture of a pianist's hand, representing the movement of striking a key, holding the key, and then releasing it.

*Example 1.* Let us give here an elementary example of the gesture of a finger of a piano player's hand, see also Figure 8.2. The gesture represents the movement of a finger going down to a key, keeping in that position for the duration of a tone, and then moving back upwards to be ready for a next movement. This gesture has three segments, which we formally relate to a gesture defined on the digraph $\Pi = \bullet \xrightarrow{d} \bullet \xrightarrow{h} \bullet \xrightarrow{u} \bullet$. The space of this gesture is three-dimensional, i.e., $\mathit{FingerSpace} = \mathbb{R}^3$, where the first is the higth above the keyboard (position, positive direction for approaching the piano), the second is the time, and third coordinate parametrizes the pitch (the coordinate of the key on the piano's keyborad). Then a finger gesture $\pi : \Pi \to \overrightarrow{FingerSpace}$ is a diagram of curves as shown in Figure 8.2.

In order to define hypergestures, we need to know, how to turn the set of gestures $\Delta@\overrightarrow{X}$ from $\Delta$ to $X$ to into a topological space. Now, the special case $\Delta = \uparrow$ means that we have the topological space $\uparrow @\overrightarrow{X} \xrightarrow{\sim} I@X$ of continuous curves $c : I \to X$ (with the above mentioned compact-open topology). The general case follows from the observation that $\Delta$ is the colimit of the following diagram $\mathcal{D}$ of digraphs: We take one arrow digraph $\uparrow_a = \uparrow$ for each arrow $a \in A_\Delta$ and one bullet digraph $\bullet_x = \bullet$ for each vertex $x \in V_\Delta$. We take as morphisms the tail or head injections $\bullet_x \to \uparrow_a$ whenever $x = t(a)$ or $x = h(a)$. Then evidently, $\Delta \xrightarrow{\sim} \text{colim}\mathcal{D}$. Therefore, the set of gestures $\Delta@\overrightarrow{X}$ is bijective to the limit $\lim \mathcal{D}@\overrightarrow{X}$ of a diagram of topological spaces $\uparrow @\overrightarrow{X} \xrightarrow{\sim} I@X$. The topology of $\Delta@\overrightarrow{X}$ is defined as the limit topology of this diagram, the space with this topology is denoted by $\Delta\overrightarrow{@}X$. In the case of a metric $d$ defining the topology of $X$, it is well known that the compact-open topology on $I@X$ coincides with the topology defined by the metric $\overrightarrow{d}(\gamma, \gamma') = \sup_t d(\gamma(t), \gamma'(t))$. And then, the topology of $\Delta\overrightarrow{@}X$ is defined by the metric $\Delta\overrightarrow{@}d(\gamma, \gamma') = \sum_{a \in A_\Delta} \overrightarrow{d}(\gamma_a, \gamma'_a)$.

This construction is functorial in both arguments: If $h : \Gamma \to \Delta, f : X \to Y$ is a couple of morphisms of digraphs and topological spaces, respectively, then the canonical map $(f, h) = \overrightarrow{f} \circ ? \circ h : \Delta\overrightarrow{@}X \to \Gamma\overrightarrow{@}Y$ is continuous and functorial in $h, f$. For example, if $h : p \to \Delta$ is the embedding of a single vertex $p$, the map $(Id_X, h) : \Delta\overrightarrow{@}X \to p\overrightarrow{@}X$ is the restriction of gestures to the point $p$. If we select a particular gesture $\xi$ in $p\overrightarrow{@}X$, i.e., a point $\xi \in X$, then the fiber $(Id_X, h)^{-1}(\xi)$ is the set of gestures sending $p$ to $\xi$. If in particular $\Delta = 1$, this fiber is the set of loops in $\xi$.

By the above, one may now repeat the gesture construction and consider the topological (and especially, the metric) space of hypergestures $\Gamma\overrightarrow{@}\Delta\overrightarrow{@}X$, hyperhypergestures $\Lambda\overrightarrow{@}\Gamma\overrightarrow{@}\Delta\overrightarrow{@}X$, etc. Notice that in particular, the space $\uparrow \overrightarrow{@} \uparrow \overrightarrow{@}X$ is the topological space $I^2@X$ of homotopies in $X$, i.e., hypergestures generalize homotopies between continuous curves.

## 8.2 The Escher Theorem

**Summary.** The Escher theorem in gesture theory states that hypergestures can be constructed from difference permutations of sequences of skeleta, but will still yield topological categories of hypergestures that are isomorphic to each other.

– $\Sigma$ –

Here is a useful proposition concerning the order, in which hypergestures are constructed:

**Proposition 1.** *If $\Gamma, \Delta$ are digraphs and $X$ is a topological space, then we have a canonical homeomorphism*

$$\Gamma \overrightarrow{@} \Delta \overrightarrow{@} X \xrightarrow{\sim} \Delta \overrightarrow{@} \Gamma \overrightarrow{@} X.$$

This results from the above fact that a digraph is the colimit of its arrows, glued together on their head and tail points. On the one hand this entails that the hypergesture space $\Gamma \overrightarrow{@} \Delta \overrightarrow{@} X$ is the limit of the spaces with arrows $\uparrow \overrightarrow{@} \Delta \overrightarrow{@} X$, which are in turn identified to the spaces of curves $I@\Delta \overrightarrow{@} X$, but these are the limit of spaces $I@I@X$ over the arrow system of $\Delta$. On the other hand, one may also first start with the limit over the arrow system of $\Delta$ and then pass to the limit over the arrow system of $\Gamma$. Thus, the two limit systems can be applied in any order, and this means that the two topological spaces in question are isomorphic by universal properties of colimits and limits. One may also use the fact that double limits exchange.

**Corollary 1.** *The action*

$$\overrightarrow{@} : \mathit{Digraph} \times \mathbf{Top} \to \mathbf{Top} : (\Gamma, X) \mapsto \Gamma \overrightarrow{@} X$$

*canonically extends to an action (denoted by the same symbol)*

$$\overrightarrow{@} : [\mathit{Digraph}] \times \mathbf{Top} \to \mathbf{Top} : (W, X) \mapsto W \overrightarrow{@} X$$

*of the free commutative monoid [Digraph], i.e., the monoid of commutative words $W = \Gamma_1 \Gamma_2 \ldots \Gamma_k$ over the alphabet Digraph of digraphs (the objects only). It is defined*[3] *inductively by $\Gamma_1 \Gamma_2 \ldots \Gamma_k \overrightarrow{@} X = \Gamma_1 \overrightarrow{@} (\Gamma_2 \ldots \Gamma_k \overrightarrow{@} X)$ and $\emptyset \overrightarrow{@} X = X$.*

*Example 2.* Referring to the finger gestures described in example 1, we want to discuss an example of piano finger hypergesture, see sigure 8.3. Before so doing, let us mention a quite intriguing statement of Renate Wieland and Jürgen Uhde in [158]:

> Die Klangberührung ist das Ziel der zusammenfassenden Geste, der Anschlag ist sozusagen die Geste in der Geste.
> (The touch of sound is the target of the embracing gesture, the keystroke is so to speak the gesture within the gesture. )

Although the precise meaning of this statement is somewhat cryptic, it could be interpreted as arguing that a gesture may be thought as being built from other gestures. Recall that a hypergesture is built from a system of homotopies of the curves that compose its vertex gestures. Now, let us take a

[3] To be precise, this action is defined up to homeomorphisms.

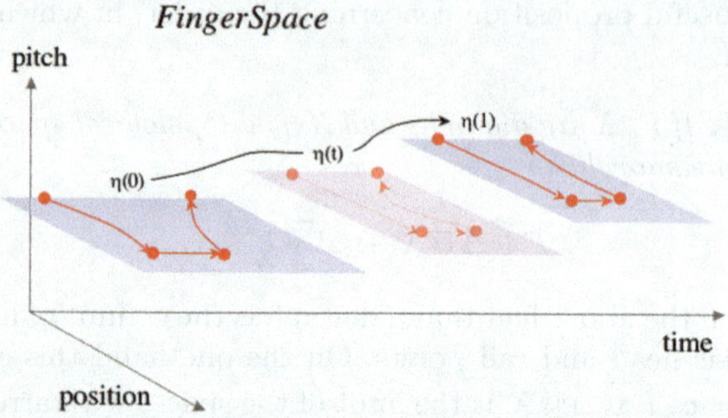


Fig. 8.3: A hypergesture of a pianist's hand, representing the homotopy curve $\eta$ between two finger gestures.

hypergesture $\eta : \uparrow \rightarrow \overrightarrow{FingerSpace}$. This means that we have a continuous curve of finger gestures $\eta(t)$, $t \in I$, from the initial finger gesture $\eta(0)$ to the final finger gesture $\eta(1)$.

Musically, this means that in the given $FingerSpace$, two gestures can be related to each other by thinking the final gesture as a result of an infinite series of intermediate gestures. So they may be connected cognitively by just deforming the original gesture to obtain the final one. This seems to be obvious in our example, since no obstruction to such a deformation is indicated. The next example shows that in a less trivial space, viz. the torus, such deformations are not always possible and from a cognitive point of view, this would lead to conceive radically different strategies in the management of gestures. In fact, the existence of hypergestures relates to the fundamental group of the underlying topological space.

Fig. 8.4: There is no hypergesture from $\eta$ to $\nu$ (left), while these two gestures are isomorphic (right).

*Example 3.* Take the torus $X = T$ and the final digraph 1, see Figure 8.4. Then, if $\eta, \nu : 1 \rightarrow \overrightarrow{T}$ are the horizontal equatorial circle curve at the origin 0 of $T$ and the vertical meridian curve at the origin, respectively, there is no hypergesture of type ↑ from $\eta$ to $\nu$, whereas the diagonal reflection on $T$ yields the morphism $Id_1 : \eta \to \nu$.

This makes clear the difference between hypergestures and morphisms: Hypergestures realize the "arrows" between vertex gestures as curves, whereas morphisms realize them by transformations between the vertex curves.

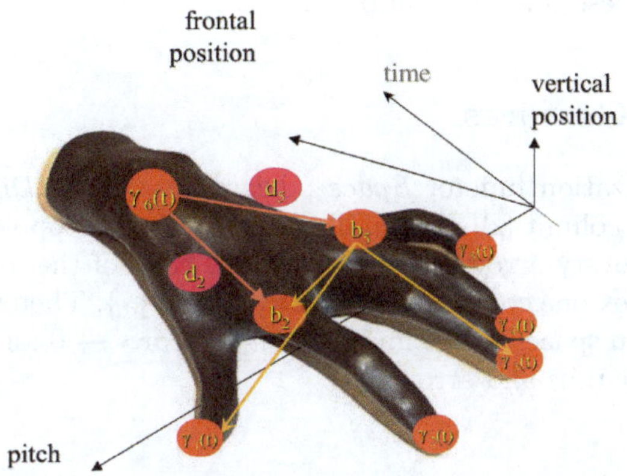


Fig. 8.5: The coordinates of a pianist's hand for its representation in a gesture space.

*Example 4.* A more involved example is provided by the gestural space of an entire hand, as investigated in [98] in the context of the simulation of a pianist's playing from a given score, under constraints from the hand's geometry as well as the physical dynamics from Newton's second law (limits of forces available from the pianist's performance). Figure 8.5 shows the hand position in time. It is given by six points $\gamma_i(t)$ in the 4-space $\mathbb{R}^4$ with three space coordinates (pitch, vertical position above keyboard, and horizontal frontal position from the key face line) and one time coordinate. The finger tips are represented by the indexes $i = 1, 2, \ldots 5$, while the carp is represented by $i = 6$. Observe that the time coordinate is not the curve coordinate $t$! A gesture's curve coordinate is an abstract parametrization of the curve in a given space, not the material time coordinate, which may also be absent.

In this example, the topological space $X$ is however subject to rather complex constraints due to the geometric properties of the hand. To begin with, the difference $\gamma_i(t) - \gamma_6(t), i \leq 5$ must comply with anatomical limitations of relative finger positions. Moreover, the relative positions of finger tips, i.e., the difference vectors $\gamma_i(t) - \gamma_j(t), i, j \leq 5$, must comply with anatomical limitations. Finally, the fingers are not allowed to intersect. This constraint may be described in a simplified way by the conservation of the orientation $u \wedge v \wedge w$ of the triple of difference vectors $(u, v, w) = (d_2 - d_5, \gamma_1(t) - \gamma_6(t) - d_5, \gamma_i(t) - \gamma_6(t) - d_5)$ for $i = 2, 3, 4, 5$. So the subspace of the a priori space $(\mathbb{R}^4)^6$ of the finger and carpus positions in space time is defined by a number of polynomial inequalities (relative position limits and orientation being described by polynomials). We denote this space by *HandSpace*.

It would be interesting to look at the fundamental group $\pi_1(HandSpace)$ and the consequences of this group for possible hand gestures. For example, the independent spatial rotation with a determined radius of two finger tips defines a torus (Cartesian product of two circles), and we are in the topologically non-trivial situation described in example 3.

## 8.3 Natural Gestures

We have spatialization functor *Space* : *Digraph* $\to$ *SpaceDigraph* as follows: We first take the colimit $|\Delta|$ of the following diagram of topological spaces (see Figure 8.6): For every arrow of $\Delta$, we take one copy of the unit line $I = [0,1]$, and for each vertex one copy of the singleton space $\{*\}$. Then we take the maps from the singleton spaces to the line copies $* \mapsto 1$ or $* \mapsto 0$ for each coincidence of arrow heads or tails and arrows.

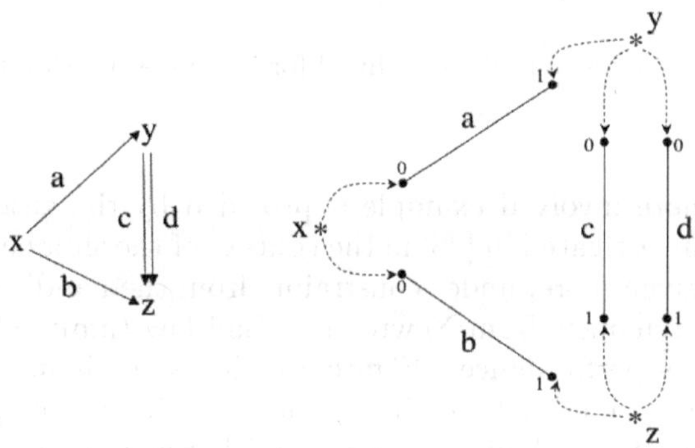


Fig. 8.6: To the right the diagram for the colimit topology derived from the digraph to the left.

The colimit topology is this: The space $|\Delta|$ is all the copies of the unit line $I$ being glued together in their common vertices, and the open sets in $|\Delta|$ are the sets intersecting in an open set for each line $I$. This space is in fact a metric space, i.e. we view it as being given the induced metric from $\mathbb{R}^{V_\Delta} \oplus \mathbb{C}^{A_\Delta}$ on the subspace consisting of basis vectors $e_v$ indexed by vertices $v \in V_\Delta$, the arrow $a \in A_\Delta$ is realized by a Cartesian product of the line $(1-s)e_t + se_h, s \in I$ for an arrow $t \xrightarrow{a} h$ times the unit circle $S_1 = \exp(2\pi s), s \in I$ in $\mathbb{C}$ indexed by the arrows. This means that we have a twisted line $((1-s)e_t + se_h, \exp(2\pi s)), s \in I$ in the direct sum $c_a(t) = \mathbb{R}e_t \oplus \mathbb{R}e_h \oplus \mathbb{C}e_a \subset \mathbb{R}^{V_\Delta} \oplus \mathbb{C}^{A_\Delta}$. We write $\vec{\Delta} = \overrightarrow{|\Delta|} =$

$I@|\Delta|$. The spirals serve two needs: (1) They disambiguate arrows which share heads and tails and (2) they make loops looking like circles and consequently spirals for "non-loops" are the most natural solution to “draw” them.

Moreover, we have natural transformation $g : \mathrm{Id}_{Digraph} \to \overrightarrow{?}$, which means that there is a gesture $g(\Delta) : \Delta \to \overrightarrow{\Delta}$ for each digraph, which varies functorially. The gesture maps the vertices $v$ of $\Delta$ to the points $g(\Delta)(v) = e_v$ of $|\Delta|$, and the arrows $t \xrightarrow{a} h$ to the curves $g(\Delta)(a) = c_a$ in $|\Delta|$ (to ease intuitive understanding, we use the representation in the metric space $\mathbb{R}^{V_\Delta} \oplus \mathbb{C}^{A_\Delta}$, but in reality we refer to the corresponding structures in the defining colimit). The gesture $g(\Delta) : \Delta \to \overrightarrow{\Delta}$ has the following universal property:

**Proposition 2.** *For any morphism $f : \Delta \to \Gamma$ of digraphs and any gesture $\gamma : \Gamma \to \overrightarrow{X}$, there is a unique continuous map $|f| : |\Delta| \to X$ such that its associated digraph morphism $\overrightarrow{f} = \overrightarrow{|f|} : \overrightarrow{\Delta} \to \overrightarrow{X}$ induces a morphism $g(f) : g(\Delta) \to \gamma$ of gestures.*

This follows by standard arguments from the universal property of the colimit topology on $|\Delta|$.

**Corollary 2.** *The map $\Delta \mapsto \overrightarrow{\Delta}$ and $f \mapsto \overrightarrow{f}$ defines a functor $\overrightarrow{?}$ : Digraph $\to$ SpaceDigraph and the gesture $g(\Delta) : \Delta \to \overrightarrow{\Delta}$ defines a natural transformation $g : Id_{\mathbf{Digraph}} \to \overrightarrow{?}$, also called the* natural gesture associated with the digraph *$\Delta$.*

**Corollary 3.** *The gesture $g(\Delta) : \Delta \to \overrightarrow{\Delta}$ defines a functor $g$ : Digraph $\to$ Gesture which is left adjoint to the projection $p$ : Gesture $\to$ Digraph, in symbols $g \dashv p$. This means that*

$$Digraph(\Delta, p(\gamma)) \xrightarrow{\sim} Gesture(g(\Delta), \gamma)$$

*is a bijection, which is functorial in both arguments $\Delta, \gamma$.*

**Impact.** This section introduces the core category of musical gestures.

– $\Theta$ –

## 8.4 Gestures in Performance

**Summary.** There are important theorists of musical performance who stress the crucial role of gestures in performance and more specifically stress the construction of musical time that goes beyond the “frozen time” in score representation.

– $\Sigma$ –

Fig. 8.7: Roger Sessions.

### 8.4.1 Roger Sessions: Questions about Musical Performance

Sessions in his book "Questions About Time" [142] stresses that awareness of movement in performance is essentially "awareness of time as a process". He claims that we experience music, and especially its time, as a gestural movement. The performer therefore is an essential contributor to the musical reality. Each performance is a renewed realization of this full existence of music. In particular, there is no "definitive" performance, it is a semiotic shifter: every performance is a new realization. And especially it is a renewal of the temporal existence of music. The performer creates musical time.

### 8.4.2 Theodor Wiesengrund Adorno's Performance Theory

In his book "Zu einer Theorie der musikalischen Reproduktion", he unfolds a very critical perspective upon the character and impact of musical notation. This is interesting as it stresses the distructive relationship of this method to musical time.

He says that notation wants "music to be forgotten", while fixing it in an objectification of the gesture. This action produces a deadly eternalization of music. Notation spatializes music and wants to create a total control, but fails as the time of music is irrevocably lost: All making music is a *recherche du temps perdu*, that famous title of Marcel Proust's epochal novel.

For Adorno, notation is an expression of a christianization of music: It beams it to eternity, it kills music as a natural phenomenon and embalms it as a spiritual entity, similar to a Platonic idea, which is an atemporal entity (see our discussion of Plato in Section 2.1). This for of survival of music presupposes killing its here and now, and detaches music from its mimetic representation.

Mimesis for Adorno is the expression of expression, the gestural representation of expressive categories.

Fig. 8.8: Theodor Wiesengrund Adorno.

The musical moment are the very arena of mimetic innervations. Music is not signifying pathos, but maps expressive categories in to their musical configuration. Performance deals with finding those encapsulated gestures to mimic them. The very concept of musical performance is the path into the empire of mimetic characters. By the visual fixation in notation, music ceases to be a gesture, it becomes an object, a mental thing.

Adorno concludes as follows:

> Correspondingly the task of the interpreter would be to consider the notes until they are transformed into original manuscripts under the insistent eye of the observer; however not as images of the author's emotion—they are also such, but only accidentally—but as the seismographic curves, which the body has left to the music in its gestural vibrations.

Summarizing, Adorno stresses the essence of gestural performance as the categorical opposite of Christian atemporal eternity. In other words: Similar to Sessions, Adorno also connects the musical time perspective to the gestural essence of performance.

### 8.4.3 Renate Wieland's and Jürgen Uhde's Theory of Pianist's Practicing

Renate Wieland and Jürgen Uhde are famous students of Adorno and have written excellent books on practising music (for pianists) *Forschendes Üben* [158].

Fig. 8.9: Renate Wieland.

In this book, the authors stress the spatial character of musical gestures, recalling that emotions are movements of gestures that bring to the surface of expressivity originally internal entities: emotions are movements brought to the external surface, this refers to Adorno's mimetic "expression of expression". They claim that therefore affects are always expressed via spatial gestural coordinates, the wording is evident as we speak about "hautiness, elevation and inclination, about greatness of mind, pettiness, about respectful and forward, etc." In other words, these authors claim that gestures are spatial categories, movements in space. They propose a geometric theory of gestures, also including gestures of gestures, what we called hypergestures, see Section 8.1. But we should pay attention no to confuse this spatiality with Adorno's spatiality in the abstraction of musical notation. Here, spatiality refers to the space of a movement, not the space of atemporal Platonic abstraction.

**Impact.** Sessions, Adorno, Uhde, and Wieland stress the essential embodied *gestural* reality of musical expressivity, including its semiotic shifter nature (Sessions) and the disruption from anti-temporal, "christanizing" notation (Adorno: "music to be forgotten in notation.")

– $\Theta$ –

### 8.4.4 Gestures in Performance of Written Music vs. Gestures in Improvisation

**Summary.** The architecture of the realization of composition and improvisation needs two basic steps: composers create the music and the performers present such creations. The sounding output of music is performance and the inner anatomy is composition/improvisation. From the performers' perspective, we exhibit two components: composition vs. improvisation. In order to perform composition and improvisation, performers must utilize gesture expressivity (see also Section 8.1). Let's first give the definition of these two musical creations.

– $\Sigma$ –

Although composition and improvisation are different, they share a number of characteristic features. We do not claim their equivalence or share an essential identity. But we shall focus on the aspects of creativity they share, and we shall try to prove that this aspect is—albeit complementary—essentially of equal weight to each other in the creation of musical works. We shall more precisely make clear that composition and improvisation turn out to be two ramifications of one and the same creative strategy: There is always composition in the genesis of improvisation and vice versa.

The types of gestures involved in these approaches to creativity and their associated time constructions appear in two endeavors. Composition and improvisation are important parts of music creation and being used or understood differently by different musical culture origins. We acknowledge that there's no pure composition or improvisation, the two spectra are intertwined with each other in any creation. But here are the definitions of composition and improvisation we offer that were built in [101].

#### 8.4.4.1 Definition of Composition

Building on the theories of Mazzola on musical creativity [101], musical composition enables us to objectivize and conserve data about musical objects. In Western music tradition, following the first neumatic attempts in the ninth century, the standard result of composition is a score [101]. The score consists of a number of note events or pauses that are inserted into a system of reference pitch, onset, duration, and loudness, together with a set of special signs to indicate articulation, loudness variations, and performance commands such as tempo and fingering. The classical score may be replaced by a set of symbols that encode other approaches to notes—symbols can be of a more graphical nature or expressions of computer language tokens, such as MIDI, denotators [94], or other symbols [101].

The compositional activity then consists of the creation of such a score or the symbols with which it is replaced. It is a remarkable fact, and we shall come back to it later in this chapter, that the original usage of notation was to generate a written trace of musical works and to describe theoretical reflections, but not to give birth to musical creations [101]. Score writing, however, changed from the documentation task to the creation of music, and it is significant that in a conservative understanding of Western music, the score is even identified within the musical work: the creation of the score being the core musical activity, while the performance of a score is reduced to a service similar to what "a MIDI slavery[4] would provide for present music technology" [101].

#### 8.4.4.2 Definition of Improvisation

A composition usually requires a set of additional actions in order to define a complete sounding musical body. This "difference set" is the improvisational part of the created music. Improvisation is, by definition, all that is not encoded in a preconceived symbolic system of commands leading to a sounding musical body [101]. In this way, improvisation is defined as the absence of something: all that is not... On the one hand, this definition looks like a negative one in that improvisation is all that is not written in those symbols. On the other hand, composition is the absence of something as well: composition is all that is not improvisation. We could also have defined composition as all that is not improvisation, where improvisation is understood positively as being the totality of all actions performed while playing the sounding music. We shall analyze this approach when dealing in more detail with improvisation. And we shall see that this alternative is somehow more complex than the chosen one. It is easier to exhibit the symbols backing a sounding musical work than to describe precisely the actions of the performing artists, hence our choice.

Although this looks like a clear conceptual setup, which could be summarized by a formula:

$$composition + improvisation = sounding\ music,$$

the details are less evident. Let us look at some basic situations. First, if the composition allows for a selection of a set of variants, then what is the status of this choice? For example, if we have a lead sheet composition in jazz with defined chord symbols, the performer would have to select one of the chords subsumed under the given symbol. Is this part of the composition or is this an improvisational action?

The critical point in this argument is that the compositional part is not well defined. If we define the composition once and for all, then the improvisational part is all but the composition, whatever composition is chosen. Of course, this definition may be done in many ways. But then, the improvisational part is improvisational with respect to the chosen composition, and this may be put into question as a critical or even an artificial choice.

Second, while the CD recording of a Beethoven sonata has virtually no improvisation at all, this is acceptable because the digital data determines

[4] "slave" is, in fact, the technical name for a MIDI-driven device.

everything here because everything is recorded digitally. Thus, there's no room for further editing except the loudness, which can be changed by the user of the CD player. In other words, the equation *composition + improvisation = sounding music* may be accepted in any case, provided the meaning of the parts is understood and interpreted in an adequate way.

We should add that the above definition seems to conceive the performative part of sounding music as an improvisational instance. This equation is, however, not the usual understanding of performance. Performers transform symbolic events of the composition into sounding events, but scholars argue about whether or not this transformation is an improvisational action. Why? Because the shaping of symbols in the performative action is not free from predetermined symbolic contents; in other words, the performers and their interpretation of the composer's score based on their training influences the performance of the piece. This means that compositional components can exist in both the written compositional and the performative action according to the given perspective.

For example, the shaping of a tempo curve might be a well-defined tempo operator, which deforms the given constant tempo of the score[5]. In fact, the tempo curve would then be subsumed under the part of the performance identified as compositional because the tempo curve follows predefined symbolic data for the construction of sounding events.

As a result, all rules of the tempo curve are compositional but there is also an improvisational part because a written composition has a "fixed ideal tempo", but performers are deforming the tempo curve in real performance time.

In conclusion, when differentiating composition and improvisation, it is crucial to recognize that these definitions are a flowing spectrum. To say in short, composition is having constructs or rules, and improvisation is everything that is not encoded with constructs and rules, but such a most basic representation is nearly impossible to achieve naturally. There are also some additional topics to consider when defining composition and improvisation. Observe that the previous definitions do not make any statements about the poietic nature of composition or improvisation. Previous definitions of composition suggested to not exclude the improvisational component [101], which brings us to the poietic nature of composition and improvisation. Poietic nature offers further insight into the creative process. Our next discourse will, therefore, focus on the creative processes giving rise to the compositional and improvisational productions.

#### 8.4.4.3 Threefold Time in Composition

The temporal characteristic of composition is threefold. First, time signifies the material parameter of symbolic events to be defined for the composition,

[5] Tempo is the inverse derivative of physical time as a function of symbolic time, see [95, Ch. 33.1.1].

i.e. the symbolic time, in units of quarter notes. Second, time is then added by the performance transformation that maps symbolic parameters to acoustic parameters. This physical time, measured in seconds, is defined by the tempo curve in elementary performance contexts[6]. As such, it becomes concrete when the composition is performed. Third, we consider the time in which the composition is being constructed. It is the time wherein thoughts about the music are deployed. Schoenberg has related this to a kind of slow-motion improvisation. The thought process of the composer constructs the composition in a mental realm that usually is much slower than the time of performance[7]. The third time where the creative compositional process unfolds could be called the logical time. In this process, the composer begins with two ingredients: the material of the composition and the rules that allow the material to be shaped and arranged in the parameter space of the symbolic events.

Although in composition the processual configuration, defined by the flow chart between material and transformational rules, looks natural, it is a very specific construction that would not work for improvisation as a real-time activity. In fact, the separation of "material boxes" from transformation rules would not work in real-time: A gesture requires total unification in the making. It is also remarkable that the real-time activity has been relativized with respect to compositional strategies in its creative potential. Evidently, composing looks more reliable, computational, and controllable than improvising. Applying a material rule leads to fewer mistakes but leads to increased uniformity within the produced music. The composer can think about the consequences of his/her decisions. The entire creation is never really risky, at least never in terms of fast improvisational decisions. Improvisational creativity challenges composers to rethink this bias to produce better results.

Improvisation is an important part of composition even though it has not been viewed that way. To truly create there must be both, a compositional and an improvisational component. The analysis of Beethoven's construction of Sonata op. 109 [66] shows this complex interaction of thinking and improvising approaches.

In the Western classical music world, improvisation is not as valued. First of all, the possibility to delay a decision when the logical time is relatively extended is an attractive alternative. Making an error by spontaneous action becomes virtually impossible. In other words, one's decisions become more and more immutable since they are issued from a reflected shaping of musical symbols.

In improvisation, such error-correction would be impossible. Perhaps this is a serious drawback of improvisation, which expresses the belief that thinking cannot be of such a high quality if it occurs as a real-time gesture. The creative

---

6 In more elaborate performances, time is given by a hierarchy of "mother and daughter tempi", see [95].

7 An exception would be John Cage's composition ASLSP, whose performance time lasts 639 years.

act in this thoughtful "offline mode" has been discussed by Michael Cherlin [20]. He classifies three types of musical memory: (1) the internal reference of a work, which one would associate with David Lewin's transformational theory [75], for example, the construction of an inverted dodecaphonic series deduced from the original series that would be played at the beginning of a piece; (2) the reference between musical works as a style-critical and tradition-oriented memory, for example, connecting the sonata form of a piece to another earlier sonata; and (3) the reference pointing to extra-musical memory, for example, the near-death experience of Schoenberg whose trace we can discover in his String Trio, op. 45. The idea behind this setup is that composition essentially is a memory process. The local and global logic of composition is based upon a referential system of memory-based delays. Cherlin's type of thought process is organized through a network of pipelines that transfer information from past to present. This network paradigm suggests that the seriousness of composition relates to its memory delay. In opposition to such memory work, improvisation would appear as a memory-less blind activity. Memory appears like the motor of thoughtfulness. Thinking can be reflecting back, or "down," onto a distant thing that objectivizes experience. "Memory-less" because it is supposed that memory access needs time (the logical time in composition). This is questionable as one may activate "muscle memory", for example, which does not need time to be activated. "Blind" is suggested since thinking needs logical time, this bias states that thinking cannot happen in real-time, that gestures, for example, are not thoughts but blind activities.

#### 8.4.4.4 Characterization of creativity in composition vs. creativity in improvisation

How could the creativity process in composition be characterized?

1. It strongly relies on the availability of symbolic musical objects as elements of a semiotic context whose syntagm unfolds in a logical time as opposed to the symbolic and eventually physical material time.
2. The creative management of such symbols heavily relies on the basic semiotic architecture, as typically described by Louis Hjelmslev [57] in terms of meta-theory, connotation, and motivation. This architecture manipulates symbols in a referential network, whose narrative is independent of material time. Accordingly, the creative core manipulation of conceptual wall inspection [101] and extension involves all of Hjelmslev's constructions.
3. The network of these symbols allows for a separation of material boxes from the connecting transformational arrows. The compositional narrativity is reflected in these networks' structures. See Figure 8.10 for the example of a dodecaphonic row (material box on left upper corner), together with three transformed rows (material boxes) that are connected by the transformations of retrograde and inversion.

How could the creativity process in **improvisation** be characterized?

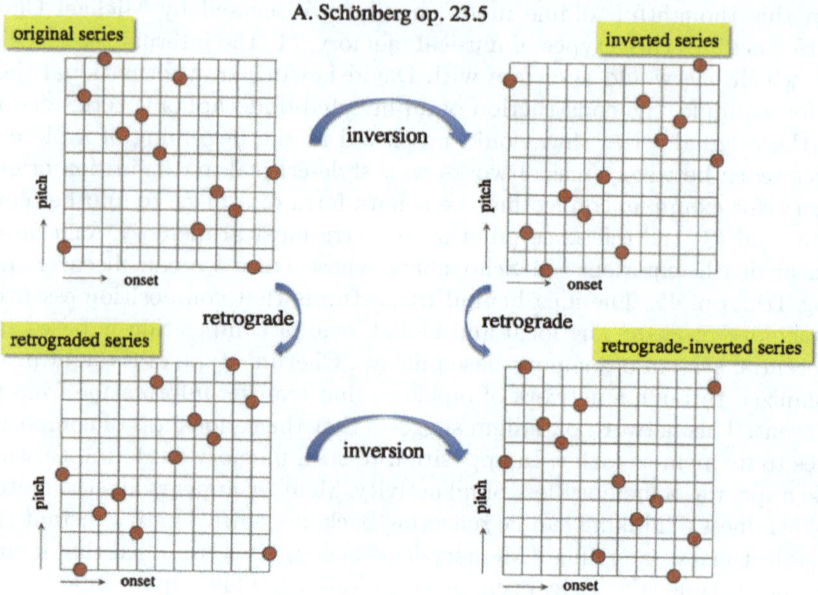


Fig. 8.10: The 12-tone row and its symmetric transformations.

1. It strongly relies on the system of gestures and hypergestures [96] that does not allow for abstract symbols and does also not enable logical time as opposed to material (physical) time. Time in improvisation is an existential dimension where gestures are living; it only exists once within gestures.
2. The creative management of gestures relies on the projection of gestures by gestural morphisms and, even more substantially, on the connectivity of gestures within hypergestures.
3. The network of gestures is always a hypergesture. There is no separation of material from its transformation, and there is no way to objectivize such gestural networks. They only live in presence. They also have no history, Eric Dolphy sais: When the music ends, the sounds are gone, and "you can never catch them again."

#### 8.4.4.5 Summary of the differences, especially the gestural aspect thereof

Composition: Compositions have three times:

- Symbolic time
- Physical time
- Logical time

Fig. 8.11: The waiter model of performance.

The gestures for this architecture of time-events are those of careers of messages, similar to gestures of servants or waiters bringing dishes to the guests (see Figure 8.11).

Improvisation: We have an interaction of gestures, in the shape of hypergestures and morphisms between gestures (refer to Section 8.1). These gestures are generators, but not careers of messages. They live in the imaginary time of the presence, no reference to the compositional times are given, see Figure 8.12.

*Remark 1.* Comment on gestural construction of musical time: Whereas in composition, time is not constructed, but taken over from the score's architecture, in improvisation, the imaginary time of artistic presence is not directly "visible" as an experience, but will be experienced as constructed from the gestural interaction. We shall discuss this constructive approach in subsequent chapters, it is in fact related to the main hypothesis of this book: musical time as a gestural construction.

#### 8.4.4.6 Performance in the European Tradition of Composition

As it has been mentioned earlier, musical notation first served multiple functions, such as documentation tasks, granting monks a level of fixity and control over the diversity of chants used in the scope of the Catholic Church and also serving as a tool for many theoretical and speculative thoughts about music, thoughts that started to abound in the Middle Ages. Neumatic notation was

Fig. 8.12: Yan & J-Sun, performing *Pianarchy*, 2018 photo credit Alice Gebura.

not absolutely fixed, allowing for a gestural representation of melodic contours and great rhythmical freedom. The evolution of notation has never stopped since. We have examples such as the early catholic chants, Arezzo's notation, Ockegem, Merulo but also more modern developments of notation that display the increasing desire for control over temporal dimensions and specific performative decisions, e.g., Bach, Mozart, Beethoven, Debussy, Bártok, Peixinho, Stockhausen.

The developments in notation led to the increasingly expanding interests of performers and theorists in seeing the score and its correct interpretation as the purest way to access the composer's intentions and hence ratify their interpretive choices and assure the validity of their executions. Two currents of thought arose from the ideals of fidelity to the score: the *Werktreue* and the *Texttreue*. These concepts refer, respectively, to the faithfulness to the composer's intentions and faithfulness to the score. The *Texttreue* tradition finds its roots in 19th century Romanticism and in the concept of Musical Work, which relegates the performer to the role of a transparent medium whose voice should be neutralized so the composer's voice can be heard [35]. Under this prism, the neutralization of the performer's voice not only disregards personal interpretive contributions coming from the musician, but also subsumes the negation of his/her physicality. Any unnecessary gestures should be prohibited in performances, since any display of physical involvement could disrupt the hierarchical valuation designed to give primacy to the score and putting the performer in a condition analogous to an invisible slavery.

The constraints imposed by the *Texttreue* ideas proved to be even more severe. German theorist and pedagogue Heinrich Schenker wrote in 1911:

> Basically, a composition does not require a performance in order to exist. Just as an imagined sound appears real in the mind, the reading of a score is sufficient to prove the existence of the composition. The mechanical realization of the work of art can thus be considered superfluous [140, p. 3].

Not only is Schenker's attitude completely dismissive of the role of performance and performer, but it is also regards performance as a menace to the purity of the composition in its ideal form, as it is represented in notation [140, p. 3]:

> Once a performance does take place, one must realize that thereby new elements are added to a complete work of art: the nature of the instrument that is being played; properties of the hall, the room, the audience; the mood of the performer, technique, et cetera. Now if the composition is to be inviolate, kept as it was prior to the performance, it must not be compromised by these elements (which after all are entirely foreign to it). In other words: those properties must not be given priority.

By assuming that the work of art is complete before the performance (as opposed to Adorno's thoughts), Schenker forbids and rejects any valuable contribution from the performer's part, limiting their contribution to a mere reproduction of the score's symbols, which are sufficient to fully carry the composer's intentions. The words of Schenker are a crystallization of a strong stream of thought that echoed in several schools of interpretation and music education during a large span of time. To offer an example, the celebrated Russian pianist Oxana Yablonskaya (who taught at the Juilliard School of Music for 25 years), when talking about her teacher Alexander Goldenweiser, one of the most celebrated pedagogues of the Russian School, declared: "He was teaching to respect what is written in music (...) and if you have personality, anyway, it will come"; showing that many performers and teachers also shared the same viewpoints of the *Texttreue*, allowing themselves, however, to insert their personality in an arcane way—what would be the artist's personality place in a code that, a priori, presents the totality of the information that are necessary to its realization?

Here the duality of traditions governing performance reveals its face—if the score provides all the information that is necessary to musical realization, what would then be the other set of rules influencing performances that grants performers the tools to fill the voids left by notation, and, maybe, display an artistic personality? Putting in different words: notation organizes different musical parameters, such as duration, dynamics (loudness), articulation in a way that is much more precise than the way it can demonstrate other aspects such as phrasing, tempo, and agogics. The shaping of a musical phrase (even if the notes are slurred, meaning that they belong to the same phrase), the length and the temporal proportions of a *ritardando* and even the pure meaning of words such as *adagio* or *presto* are much less obvious, for example, than the realization of rhythm and pitch. The very shaping of time is poorly represented in traditional notation. *Rallentandi* and *accelerandi* have no precise and uniform way of execution. *Ritardandi* that are commonly associated to phrase endings can hardly be controlled by notation—Alban Berg's Piano Sonata Op. 1 exemplifies a dramatic attempt on trying to control this aspect. Even metronome marks, an important technological development of the beginning of the 19th century and highly celebrated by Beethoven, have proven not to be absolutely precise: Schumann's infamous metronomes marks, claimed to be impossible by many pianists, and the discussion on whether metronome marks should be followed through an entire piece or are just an approximation to the ideal tempo as projected in the composer's mental reality are some examples among many other concerning the usage of the metronome.

That is where the oral tradition comes into action. Publicly, one proclaims the faithfulness to the composer (and the text), but privately, one obeys the rules of oral tradition [35]. Liszt recognized the importance of the performer, for example, in the preface of his Symphonic Poem *Ce qu'on entend sur la montagne*—he attributed the completion of his work to the performer, since important aspects of his intentions could not be written down on the score [64, p. 7]. Although adhering to *Werktreue* ideas, Liszt recognizes a dependence on the performance to complete his works. The entry on Liszt in [38] offers another insight about Liszt's ideas on interpretation:

> In the 1830s Liszt developed some unconventional marks of expression. The impulse to do so arose from a youthful desire to control every aspect of interpretation, especially tempo rubato. The 1838 version of the *Transcendental Studies* offers an abundance of such devices. In later years, when he revised much of his early output, Liszt dropped them, presumably because he felt that such matters are best left to each individual player. Their chief interest today is that they tell us how Liszt himself might have interpreted his own music.

Liszt apparently gave up on trying to control how his music should be performed in every minimal detail, investing the performer with the power to take part actively as creators of meaning and musical content when performing his works. This performative activity is not, of course, free from constraints. Romanticism, as opposed to the Baroque period, for example, did not explicitly set rules for the interpretation of musical works. What one could see on the 19th century was the increasing importance of the oral tradition as an important factor for the success and coherence of a performance. Perceiving the impossibility or the limitations imposed by the notation on the exact transfer of precise performative details, a tradition based on information received from teachers and the contact with different performers and their executions flourished.

Rationales that went beyond the limits of what could be notated on the score gained more importance as technological developments were implemented in the instruments, what also opened a path for technical developments on the performers' side. The piano music of Beethoven, for instance, was marked by a physicality that distinguished it from the works of other composers that were performed in Vienna, such as Mozart or Eberl [34]. Gestural rationales (considerably hard to emblazon on traditional notation) played a determinant role on how these works should be performed according to rules that would still make them feel recognizable as pertaining to the style of a composer, or conveying an acceptable message from the point of view of rationales such as drama narrative. One example is Glenn Gould's interpretation of Beethoven's Sonata Op. 57, *Appassionata*.

The 'misinterpretation' of Beethoven's text by Gould comes by the pianist's unawareness or negation of an inbuilt performance grammar that is not engraved in the score but stems from the performance tradition as such, as an element of rhetorical communication that transcends the code [101]. This void in Beethoven's notation finally meets its uniformization in tradition: one would say that is easy to notice, assuming the necessity of a gestural component to the understanding of this passage, that the huge dominant seventh chord in measures 14 and 15 requires a temporal modulation and increased dynamics in order to convey the drama of the cadence that precedes the repetition of the piece's first phrase. The unexpectedness of the forte burst occurring in these two measures seems to be magnified by tradition or by an underlying gestural information that is hidden in the score. Beethoven's text mandates this passage to be executed forte and a tempo. The question is whether the dramatic character is implicit in the score structure or has been added by historical standards [101], since dynamics louder than forte and an increase in tempo that represents more than the a tempo indication placed by Beethoven are commonly associated with the performance of this gesture.

Gould's interpretation of Beethoven's *Appassionata* gives us a unique perspective of this piece. It is hard to avoid making connections with Schenker's *Texttreue* perspective. By applying to a Beethoven's piece the same millimetric procedures that made Gould's renditions of Bach so famous (that are also the mark of the *Texttreue*) the interpreter seems to disrupt the stablished performative guidelines of the piece. What comes out is a result that, although in perfect accordance with the explicit information of the text (*forte* and *a tempo*), defies all conventions of what regular audiences would associate with a standard performance of the piece.

Glenn Gould's performance of *Appassionata* is an allegorical example of the way gestural manipulation can create and influence temporal constructions. The Canadian pianist illustrates how the comprehension or the manipulation of gestural rationales that go beyond the overt rules of notation, and that in

Beethoven's Op. 57 are commonly related to the piece's implied drama, are capital to the shaping of time.

#### 8.4.4.7 Uniformization by Recordings

Be this syntax of gestural components a part of the idiom of a composer, implicit in the score structure as a repertoire of gestural formulas used by several composers, or a result of layers of different historical standards, the semantics of interpretation has had, over the past 100 years, one more system of models to contribute to its uniformization: a whole tradition based on recordings.

Under Taruskin's perspective [154], performances could also be considered as a part of oral tradition. But the massive amount of available interpretations of the same work that are sometimes just a click away from performers imposes on musicians constraints that go far beyond the advices passed from teachers to students or an expressive vocabulary that is tacitly learned as performers attended to concerts or have access to a limited number of performances recorded on CDs or LPs. The quantity of recordings available to audiences and performers soared with the implementation of the online video-sharing platform YouTube and the other means of digital distribution. Recordings of classical musical have existed for more than 120 years, however. The German label *Deutsche Grammophon* started its activities in 1898, event which also launched the advent of distribution of recorded music by means of discs. The same label was also responsible for starting the commercial distribution of Compact Discs in the 1980's, event responsible for an unprecedented increase in the availability of different recordings.

A question arises from this whole body of information that is abundantly available to performers, be they renowned professionals or students: whether the cumbersome quantity of performances so easily accessible might be changing the ways the semantics of music interpretation are being perceived and reproduced. As connectivity increases, accessibility to information becomes almost instantaneous, and the whole corpus of human knowledge grows in volume and shareability at an unprecedented pace (here included recordings of classical music), it would be foolish to believe that principles of musical execution and interpretation would come unscathed trough all these changings. Of course, the access to different recording can be positive: it enables performers to reflect on their own learning and creative processes, encourages them to potentially expand their expressive vocabulary and to have critical attitudes towards their own interpretive choices.

The demands for commercial success and massive distribution of recordings have also changed the way records are listened. Eventual mistakes like note errors can easily be corrected with computational softwares (which was not possible, for instance, when Alfred Cortot was making his recordings), and most devices that are used to reproduce music do not usually have the capacity to accurately transmit all the dimensions and complexity of a music work.

Loudness can be easily adjusted in every device, thus creating one more improvisational dimension on the recorded music. The element whose immutability is most preserved in the experience of listening to recorded music is time.

The perception of the "time of others", the time that is perceived when we listen to recordings, can be capital when performers create their "own time". Freitas's [42] findings point out to the positive influences brought by the audition of recordings by three undergraduate piano students, when analyzing the timing of their performances. It is especially interesting the fact that one of the students, despite having overtly refused the influence of the models over his performances, tacitly absorbed interpretive information from one of the audio recordings to which he had been exposed, concluding that "in the lack of preestablished interpretive ideas, he indeed absorbed, in an unconscious manner, some interpretive ideas from the model" [42, p. 91]. The fact that performers inadvertently can copy ideas from auditory models to which they are so easily exposed when they still lack interpretive ideas of their own relatively to the pieces they are practicing could actually hamper the development of a personal artistic voice, leading to a massification of standards of musical interpretation.

Uniformization can also be a natural symptom of the mass distribution of music. The positive side of this system of distribution is that the access the audiences have to music increases. The other side of the coin is the uniformization that this very process entails. It is not unusual, when talking about musicians of the beginning of the 20th century, that we hear commentaries about how unusual their interpretations were, about temporal freedoms, about either playing too fast or too slow, that their approaches were too personal. The risk brought by the omnipresence of recordings is that they can become more and more isomorphic, as a way to grant they commercial success and acceptability[8], and hence influence this and the future generations of musicians to search for ideals of performance that include its uniformization, ultimately leading them astray from the quest of a personal artistic voice.

The ambivalence between a personal artistic voice and the impositions of a market that creates and sells homogeneity reflects on the other question we had sketched before: the creation of "one's time" versus the creation of the "time of others". Free jazz serves as the best example for the experience of creating new time, a time that is free from preconceived walls. The characterization of free jazz brought in [99], emphasizes the responsibilities the musicians have when creating this kind of music, including the temporal dimension [99, p. 7-8], emphasis added:

> Free jazz is that jazz, *where the musicians take their full responsibility of what is being played.* (...) *The musicians do not follow an a priori type of script*; they generate the music as if partaking in a dynamic and sophisticated game. (...) In such a game in the making, *the performers are necessarily constantly shaping the body of musical time.*

[8] Confirming the sad fact that listening reduces to reproducing one's earlier experience: cognition as recognition.

Free jazz is then a strong example of how musicians are directly involved with the creation of time when they perform, not only because gestures in free jazz are newly created in every performance, but also due to the absence of a preconceived code dictating how the performance should evolve.

One has little to no expectations about what will happen in a free jazz performance since all the elements that make up the performance are built from an in-time negotiation among the performers. As opposed to a free jazz, which is much more balanced toward improvisation elements, we have the performance of score-based, Western classical music. All the before-mentioned facts have impacts over the improvisational components of classical music: the efforts made by composers in filling the code with more performative directions; the influence of oral traditions; and the effect of the massification of recordings. It is very easy nowadays for musicians to recreate the "time of others" instead of creating "their own", so abundant are the models and standards to follow, which causes the performance of classical music to be much more inclined nowadays towards its compositional parts then it was a century ago.

**Impact.** This discussion of the spectrum between composition and improvisation in performance proves (among others) that the ownership of time is a crucial characteristic that differentiates these two polar forces of musical creativity. And it also proves that this temporal ownership is expressed by the gestural mediation of time, in fact a delicate ontological category that dramatically transcends score-driven tyrannies.

– $\Theta$ –

### 8.4.5 Cecil Taylor's *Burning Poles*

Fig. 8.13: Cecil Taylor in the DVD "Burning Poles", 1991.

#### 8.4.5.1 Time as a Central Existential Category

Time management is essential for life, and it is an art in musical performance, where it refers to the artful skill of when to play what music elements. In a classical music ensemble, the conductor is the "master of time". In contrast, a creative jazz ensemble allows individuals to react (with each other's sound) in their own timing but in a collective effort. A classical ensemble designates a lead player and/or a conductor to be the sole leader, and the structure (the score) needs to be studied and rehearsed, see Figure 8.14. In Cecil Taylor's gestural performance, everyone manages their timing, but in a collective effort, allowing individuals to react to each other's sound and participate from a place of individuality, every musician can be the leader, and being a follower depends on the particular situation; the structure is loose but still present. This interaction is not the tool to achieve another goal, it expresses an autonomous artistic self.

Fig. 8.14: Rehearsal of Yan Pang's *Philia in Melody* [167].

A lead sheet is the standard way a jazz composition is presented in a written score form. It contains only the essential melodic movements, together with a sequence of chord symbols, the chord changes, written in standard musical notation. By looking at the lead sheet of a composition, one immediately recognizes the chord changes (this is the official term for the harmonic sequence of a lead sheet). Coltrane's *Giant Steps* (1960) "jumps around the entire harmonic spectrum" very fast and breaks from the traditional harmonic progression on a classical jazz lead sheet [102, pp. 181-183].

#### 8.4.5.2 The Lonely Time

The autonomous interaction is not a tool for another goal, it is a satisfaction that creates the group's own time as an ultimate goal. This interaction is a collaborative effort, which is opposed to the "lonely time of the individual male genius". This traditional individual time, especially in European tradition (including the avant-garde), is only a tool to achieve another goal, time is never the center of one's (re-)creative efforts.

#### 8.4.5.3 Body Interaction

Fig. 8.15: Cecil Taylor's body interaction.

In this autonomous construction of musical time, the musician's the body is no longer interacting with the instrumental body as a sound production machine, but as a body partner. Similar to time, the time generating gestural interaction of bodies becomes an autonomous artistic dimension beyond simple machines.

In 1991, Cecil Taylor recorded the video *Burning Poles* with William Parker on flute and bass, André Martinez on drums and percussion, and Tony Oxley on drums, see Figures 8.13, 8.15. Here is our description from [99, p.76]:

> Taylor appears as a magic cantor and dancer; he floats about his open grand piano, singing and reciting cryptic incantations while scratching and plucking the internal strings of the piano, or hitting the strings with a soft mallet. The music and performance is a percussive shaping of the body of time in a tensed trajectory of gestures, and is an excellent illustration of Châtelet's circumscription of "gesture". Taylor's

performance radically differs from the music as constructed in *Unit Structures*—there are no structures to be broken, no starting and terminal points of functions, no empty time spaces where time shards are imbricated. Taylor invokes a big, smoothly flowing gestural organism that feeds its identity by its incessant unfolding and caressing of time and space. The piano's strings are not merely objects for sound production, but extensions of Taylor's aesthetic of creating and sculpting innervated space-time. Although some of the lyrics of Taylor's *Sprechgesang* may convey normal English words, their extra musical meaning is completely irrelevant, as any kind of localized or preconceived meaning is irrelevant, the gestures are pulsating in a germinal ontology of pure making.

Taylor is creating pressure on an instrument to make the desired sounds. One achieves this by adjusting the posture and angle on the seat. The rest of the body is involved, not just the fingertips. "It was always the whole body that emitted sound: instrument and fingers, bend. Your ass is in what you sing. Dedicated to the movement of hips, dedicated by that movement, the harmolodically rhythmic body" [115, p. 39-40]. Moten's "Your ass is what you sing" is a way of saying your whole body and soul are engaged in performing. Art cannot be separated from the artist any more than your ass can be separated from you.

#### 8.4.5.4 Silence for Cecil Taylor

Cecil Taylor stresses dance/gestures similarly to physical travel through time, whereas classical music deals with "imaginary time." (see also Chapter 12.1) Taylor's jazz style is a radical break with classical musical traditions, including traditional jazz when the form was created. And Taylor's free jazz[9] is a double rupture and a far-reaching innovation of the construction musical presence, that resulted in "gestural dance" [99, p. 149]. Taylor attempted to "imitate a dancer's leaps and movements" in playing the piano and performed a dance with his hands that emphasized physical performance and made visible the Black body that has always been present in American society [99, p. 149].

#### 8.4.5.5 Performance of Cecil Taylor, Black Body Revolution

Performance of Cecil Taylor is a unique interaction of human and piano bodies. He sets forth the idea of the black body's prominence as opposed to the Christian negation of the body as a sinful necessity. His dialogue with the piano's body is an emancipated action that expresses the full-fledged role of bodies in the artistic creative process. Cecil Taylor critiqued the absence of the black body by calling attention to, or centering, his body in live musical performance.

[9] See [99, p. 7-8] for a definition of free jazz, our citation was given in Section 8.4.4.7.

#### 8.4.5.6 Dystopia

Rather than submitting to the reality and social norms that see black bodies as socially and economically stagnant, Taylor moved his body and made himself visible. From this place of hope through innovative movement, the future of jazz expands. This visibility criticized the society by calling attention to the physical body, which allowed Taylor to comment on the absence of the black frame on stage in performances. Moments of freedom on stage provide hope for the future because the audience imagined themselves moving freely along with Taylor.

The Avant-Garde movement made revolutionary artists, and those artists made revolutionary history. In his book *In the Break* [115], Fred Moten addresses blackness and black music as a specific mode of inquiry. The object is jazz as the avant-garde. We aim at drawing out influential concepts early to the mid-20th century in ways that inform and give depth to research on music, particularly avant-gardism, and leading jazz artist Cecil Taylor was avant-garde for his own time; he is credited with starting the free jazz style. Taylor invented piano playing as a percussive instrument, hitting the piano keys like drums instead of a melodic line. He moved more and more toward playing the rhythm (instead of chord changes and melodies) starting in 1955, when he initiated his revolutionary style at the New England Conservatory. Taylor still followed the rhythm of traditional jazz. In 1966, Taylor destroyed the rhythm, making his music sound like there was no rhythm, just musical elements flowing around (*Cecil Taylor—Unit Structures*). He found his own rhythm, which is a fast triple rhythm, unmeasured, and more of a pulse. He explored this technique until the end of his performing life. Taylor's music is based on the embodied energy of his playing. The fact that his playing is transformed/created by the energy of his body, not just by notes and rhythm, is what makes Taylor avant-garde.

Taylor was one of the first jazz musicians to deliberately incorporate body gestures into his performance. In Taylor's *Burning Poles* [155], he dances around the piano. He avoids playing the central part of the instrument, which are the actual keys. It is a dance of his hands, but also with his entire body, which I argue is the critical dystopian theme in his work. Critical dystopia offers transformative elements and a hope for redemption of the human race. For Taylor, a transformative part is that the previously absent body can be brought into societal awareness by performance. '*Burning Poles* is a representation of this possibility and demonstrates how it can be accomplished. The redemptive quality of humanizing the dehumanized invisible body benefits everyone if the perfect portion of Taylor's vision is to bring equality to reality.

**Impact.** Taylor's form of critical dystopia and its transformative elements expressed through gestural dance transformed how free jazz was played. This creative process of music tradition, transformation, and production is why we regard jazz as a more radical critique of contemporary society than the historical avant-garde. Taylor's works are based on the great musical heritage of jazz, based on critical dystopian themes. They have a transformative element that promises a better future, but there is hard work to be done first. Taylor's body reminds observers of what is invisible, the Black body, with his body, and tells us that we can do better, just put your ass into it.

– $\Theta$ –

# 9

# Kramer's Time Concepts

**Summary.** In what follows, we want to propose some interpretations of Kramer's conceptual variety in the framework of mathematical gesture theory. The main purpose of this action is not to put his approaches into question, but to show how strongly his conceptual sketches are congruent with mathematical conceptualization.

– $\Sigma$ –

Fig. 9.1: Jonathan Kramer proposes four categories of time: Multiply-directed, gestural, moment, and vertical.

In Jonathan D. Kramer's impressive account on time in music [72], we are given a rich spectrum of different time concepts. Kramer however stresses that he does not present strict definitions, he states [72, p. 6]: " *Time* must ultimately be taken as undefinable." He then concludes [72, p. 8]: "They (the categories of time) are more suggestions for listening than rigorous theoretical formulations." Nevertheless, Kramer proposes quite precise conceptualizations, also traced in a clear Glossary at the end of the book.



G. Mazzola et al., *Making Musical Time*, Computational Music Science,
https://doi.org/10.1007/978-3-030-85629-8_9

## 9.1 Kramer's Time Variety

**Summary.** It is important to understand how Kramer positions musical time as an ontological locus.

– $\Sigma$ –

Before we delve into Kramer's time categories, it is important to understand how he positions musical time as an ontological locus. He writes [72, p. 6]: "In the following chapters I postulate many types of musical time. To do so becomes possible once I accept the notion that music creates time. (...) Time itself can (be made) move, or refuse to move, in more than one 'direction': not an objective time out there, beyond ourselves, but the very personal time created within us as we listen deeply to music." This constructivist position is essential in our approach, too. Kramer distinguishes "ordinary" physical time from its musical counterpart [72, p. 7]: "The difference between ordinary lived time and musical time is, according to Clifton [28], the difference "between the time a piece takes and the time which a piece presents or evokes."(...) All the species of time discussed in this book are experienced and understood *simultaneously* with ordinary time."

Kramer proposes four categories of time: Multiply-directed, gestural, moment, and vertical. Here are the definitions in his glossary:

- **gestural time:** species of multiply-directed musical time in which the conventional meaning of gestures (beginnings, endings, structural upbeats, etc.), rather than the literal successions of events, determine the logic of continuity.
- **moment time:** temporal continuum of a moment-form composition
- **multiply-directed (linear) time:** temporal continuum in which progression is seemingly in several directions at once
- **vertical time:** temporal continuum of the unchanging, in which there are no separate events and in which everything seems part of an eternal present

In the following sections, we discuss Kramer's time concepts from his glossary and also from other locations where he defines time qualities.

## 9.2 Gestural Time

Gestural time is a special multiply-directed time, where gestural determinants override the literal succession of events. If we insert our definition of a gesture, it has its parametrization, which is essentially a linearization on each arrow of the defining digraph. We would define gestural time as this type of parametrization. The gestural parameter is not a physical time, the body of a gesture only would have physical time as a parameter in a space-time. If a hand moves in physical space-time, its gestural parametrization is different from the time coordinates of the gesture's curves. Moreover, if we deal with hypergestures, the Escher theorem yields a variety of gestural parameters. In fact, if a gesture $g$ is in the space $\Delta_2@\Delta_1@X$, the parameters of $\Delta_2$ can also be replaced by the parameters of $g$, when interpreted as living in the homeomorphic space $\Delta_1@\Delta_2@X$. This can be read as an instance of a multiply-directed time: gestural parameters are multiply-directed. Therefore we interpret *Kramer's gestural time as the parameter set of a gesture that is given by its defining digraph.*

## 9.3 Moment Time

We shall not discuss this type of musical time for two reasons: to begin with, it plays a minor role in Kramer's theory, and second, it is absorbed in the central category of vertical time.

## 9.4 Linear/Absolute Time

Linear/Absolute time is the traditional external time frame imposed by the metronome and referring to physical reality. This is the opposite of Kramer's focus, it is not a time that can musically be constructed, it is absolute, disjoint from musical creativity.

## 9.5 Multiply-directed (Linear) Time

Multiply-directed time will not be defined in general here, but we have already given its interpretation for the case of gestural time. It means that the gestural parameters are multiple (even on one determined digraph, we have several arrows with their associated parameters), and this in a dramatic variety when considering hypergestures and their variants that are defined by the Escher theorem.

## 9.6 Vertical Time

This is the musical time par excellence, very difficult to conceive, but it is the type of time we want to interpret in mathematical gesture theory. Let us give some references to Kramer's conceptual sketches. Two citations seem central.

First [72, p. 55]: "But some new works show that phrase structure is not a necessary component of music. The result is a single present stretched out into an enormous duration, a potentially infinite 'now' that nonetheless feels like an instant. In music without phrases, without temporal articulation, with total consistency, whatever structure is in the music exists between simultaneous layers of sound, not between successive gestures. Thus, I call the time sense invoked by such music *vertical*."

The second citation describes the way one listens to vertical music [72, p. 57]: "Listening to a vertical musical composition can be like looking at a piece of sculpture. When we view the sculpture, we determine for ourselves the pacing of our experience: We are free to walk around the piece, view it from many angles, back and view the whole, contemplate the relationship between the piece and the space in which we see it, close our eyes and remember, leave the room when we wish, and return for further viewings."

This means that vertical time creates an imaginary space where time takes the shape of a spatial category, with multiple dimensions that are perceived as if we were observing a sculpture. This perceptual work is the 'performance' of an interpretative trajectory of one's senses/eyes which move on the sculptural surface similar to what Paul Klee described as the trajectory of one's eyes when looking at a visual object [67, p. 67].

This category is a very difficult one, and we could not identify the mechanism that would generate vertical time in Kramer's book. It seems however plausible that from the above description of the imaginary space, vertical time is generated by a collaborative effort of gestural components. The dynamic perception of a sculpture is in fact the integral of a variety of gestural trajectories. In other words, we could argue that vertical time is the result of a deep interaction of gestural time perspectives.

## 9.7 The Twentieth Century Technological Time Revolution

**Summary.** Kramer discussed the role of technology in the changes of musical time management, in composition and perception.

– $\Sigma$ –

We should start this section with a caveat: Kramer's book was published in 1988, his knowledge about and the state of musical technology could not include the technological revolutions initiated by Apple's iPhone, iPod, iPad, and the internet's introduction of an omnipresent high-performance music distribution via YouTube, AppleMusic, Spotify, etc.

In Kramer's chapter "The Impact of Technology", he discusses the effects of music technology, from early reproduction on disks to tapes, musique concrète, Glenn Gould's extensive tape slicing activity for his Bach recordings, Karlheinz Stockhausen's understanding of the electronic music potentials while working with Herbert Eimert in his studio in Cologne around 1955, and Pierre Boulez's understanding of the fine-grained management options with electronic equipments that he had access to at the IRCAM in Paris.

Kramer's conclusions are that the possibility of indefinite reproduction of performed composition has changed to identity of music altogether, one may compose structures that cannot be understood from a one-time listening, but they are conceived to be understood after a multiple listening. He observes that the difficulty of a number of 20th century compositions prevents performance for many years. We should add that already Beethoven's Sonata op. 106, "Hammerklavier", composed in 1818, could not be performed before Franz Liszt premiered it in Paris in 1836, 18 years later.

The technological control of time also enables the composition of a temporal anatomy that by far exceeds traditional granularity. Kramer then discusses the conflict between machine-generated rhythms and human creation or interpretation of rhythms. Kramer also refers to Ingmar Bengtsson's and Alf Gabrielsson's psychological investigations of musical vs. performance complexity.

When he wrote his book, electronic music technology was not capable of shaping delicate and hierarchical time structures. This has been achieved around 1989, see [92]: there is software to construct complex tempo hierarchies, see our discussion of tempo hierarchies in Section 7.2. Nevertheless, human time constructions can be understood from the vertical time procedure which is applied when a competent performer shapes the body of time. Recall that conductor Serge Celibidache refused to record his performances because he rightly claimed that each performance is unique, it is a semiotic shifter, which changes its content with every single usage of the scores symbols.

Kramers final statement in this chapter is visionary: he argues that modern technology increasingly enables the fusion of the roles of composers and consumers of music. This trend would converge into machines that could be called PCs: Personal Composers. A consequence of such a development would clearly be that the construction of musical time would be shared between composers, engineers, and consumers, time would become an instance of a technologically enhanced *distributed identity*, see Chapter 10 for this concept.

**Impact.** Kramer envisioned an impressive variety of time concepts that included the composer, interpreter, and listener. His most advanced conceptual sketch is *vertical time*. Despite its sketchy conceptual precision (also explicitly notices by Kramer), we now understand its central role in the construction of musical time, because it is the vertical time, where the human creative power enables a radical opposition to the passivity of chronos in physics.

– $\Theta$ –

# 10

# Distributed Identity in Music

**Summary.** We introduce the concept of a distributed identity in musical performance. This is about the construction of a musical identity that is issued from a collaboration, where eventually the music starts playing you and not you the music, a higher axis of musical identification in the making of musical reality and in particular: time.

– $\Sigma$ –

The idea of a *distributed identity in musical performance* was first presented in Mazzola's free jazz book [99, Chapter 11]. In that context, performance plays a dominant role since there is no score or lead sheet in free jazz. The music of free jazz is generated in the making, and with the full responsibility of the musicians, a characteristic attribute of free jazz and more generally free improvisation.

Therefore, the identity of the created free jazz music cannot be reduced to or grounded on a mental template, a radical difference to score-based classical European music. Of course, when free jazz was created there were a number of unifying motivations, such as the political and cultural revolution against suppression of human rights of African Americans and against the "whitened jazz" norms, which would not allow musicians to play the music, but would enforce them to "play the background", as Ornette Coleman phrased it in his remarkable liner notes to the*Free Jazz* album [29]. But these vectors did not define the music as such, they rather created a social atmosphere wherein music would emerge with new forces.

This revolutionary opening movement was not a positive shaping force but rather a negation of grown and enforced limitations. In other words, the question of what to play, and how to do it, was not answered by the big step outside those norms. In particular, it became a core topic to create this new music's identity, to approach the question of "what it is", what could be its identity *qua* music, and not only *qua* revolution against dead and destructive norms.



G. Mazzola et al., *Making Musical Time*, Computational Music Science,
https://doi.org/10.1007/978-3-030-85629-8_10

## 10.1 Collaboration in Free Jazz

**Summary.**We discuss the idea of a *distributed identity* to understand the construction of an identity in musical collaboration.

– $\Sigma$ –

One first and decisive step towards a new identity in free jazz was the fact that the elimination of a given musical reference as defined in scores and lead sheets left the musicians alone, and thereby enforced them to relate to each other in a reference-free collaboration. The working musicians became their only reference, which meant that you could only judge your improvisation by the action and reaction of yourself and your fellow musicians. Even the audience would become less relevant since they were still biased by tradition and standards, and often, free jazz concerts had more musicians onstage than music lovers in the audience.

The collaborative efforts therefore became the most important value system, a kind of rotational dynamics that could be realized by the incessant exchange of musical messages. This system of musical values got its stability from the living interaction, and not from any external template.

This logic implied that the *musical* identification of such an art had to refer to the collaborative efforts, and nothing else. In particular, the creation of such an identity was necessarily rooted in *all* participant musicians, it had to be a *distributed identity.* This exclusive reference to the engaged musicians would in particular not admit any imported external temporal reference. The traditional starting jazz sentence "a one, a two, a one two three four...", setting up a beat reference, was eliminated, much as the usage of a conductor's directives or a metronome's tic tac. *Time in free jazz had to be created from scratch in their collaborative interaction.*

The nature and ontology of a distributed identity is the central problem when we want to understand how the collaborative interaction works. We don't have symbolic templates anymore, no notes, bar lines, metronomes, etc. Free jazz musicians are forced to interact with their musiking bodies, to produce embodied gestures and to set up a dialogue of gestures. Nothing else is available. Therefore an elaborate analysis of a distributed identity must rely on its gestural ontology. The distributed identity must be generated from the underlying gestural interaction. It is not astonishing that free jazz pianist Cecil Taylor stated that in traditional Western music, the body plays no role, see also Section 8.4.5.

## 10.2 When is Free Jazz Successful?

**Summary.** This section covers the topic of success in free jazz. This situation is extreme in the sense of an artistic activity that is separate from any score reference and therefore completely relies on what happens based upon pure collaboration.

– Σ –

The question of when free jazz is musically—and not commercially—successful is a new problem since nobody can now refer to external norms or requirements, whose compliance would 'prove' one's success. The question refers to the ideation of gestural interaction in the creation of the distributed identity.

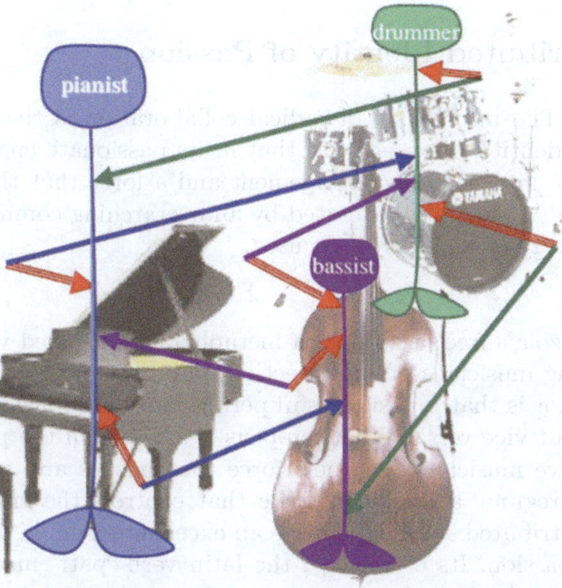


Fig. 10.1: The gestural interaction of a jazz trio. ©Guerino Mazzola

The interaction of musical gestures has two components: a set of gestures, and a set of relations between gestures, i.e., the perspectives between gestures, how one gesture views another gesture. In our presentation of gesture theory, we have exhibited such perspectives, namely morphisms $f : \gamma \rightarrow \delta$ from gesture $\gamma$ to gesture $\delta$. A morphism is an expression of how to look at gesture $\delta$ from the point of view of gesture $\gamma$. More precisely, the gestures, which are involved in a musical perspective are very often hypergestures, gestures of gestures of gestures,... For example, the drummer's gesture may be a loudness gesture of snare gestures of left and right hand gestures. But we know from the Escher theorem (see Section 8.2) that hypergestures can be viewed according to different permutational choices of their components; the drummer's hypergesture can also be interpreted as a left and right hand gesture of a loudness gesture of a snare gesture. This inerpretational variability is part of the (hyper)gestural interaction, it informs us about the perspective which is taken for a given morphism. In Figure 10.1, we show the gestural interaction among the piano, bass, and drums of a trio. The Escher permutation of the domain of each morphism is symbolized by the level of the morphism's vertical domain line, in red color, the Escher permutation of the codomain is shown as the second level that is addressed by the morphism arrow's head.

From this diagram of (hyper)gestures, the success or failure of the gestural interaction must be deduced. This is the conclusion, which was drawn in Mazzola's free jazz book.

## 10.3 Distributed Identity of Passion

**Summary.** The mechanism of radical collaboration is the construction of a distributed identity, of an activity that in its passionate engagement is at the same time an intense active engagement and a force that that is subject to a "higher force", a passivity generated by an overarching command. The double-sided etymology of "passion" is focused.

– $\Sigma$ –

In Mazzola's free jazz book, a metaphor was invoked which is quite well known among musicians of any direction, improvisers and classical ones alike. This metaphor is that in a successful performance the music starts playing the musicians, not vice versa. What happens is a flip from the passionate engagement of active musicians to a new force that pushes and pulls them into an overarching regime, a dominant force that controls the entire orchestra and gives it a distributed stability. This is an excellent example of the double-faced concept of passion. Its etymology, the latin verb "pati" means to suffer, it is a passive attitude. But the modern meaning is that passion expresses an extremely active engagement. When the music starts playing the musicians, the experience exatly this flipping movement: their active passion mutates to the passive "pati", they are thrown into a regime of a 'higher' unifying force.

In a more embodied interpretation of this transfiguration, one could view the collaborative exchange of musical gestures as a rotational dynamics that creates a gyroscopic axis, which now has the musicians rotating around itself. The success of this passionate collective effort is the creation of a stable gyroscopic axis. This was the analysis given in Mazzola's free jazz book. The example of Archie Shepp's album *Coral Rock* [144] in Section 114 of that book illustrates this gyroscopic effect of a distributed identity in free jazz.

*The progress we present the first time in this book deals with a precise mathematical setup to conceive the distributed identity. We shall discuss this topic in Chapter 11. In particular, we shall introduce the technique, which leads to our model of a gestural construction of musical time.*

## 10.4 Distributed Identity outside of Free Jazz

**Summary.** This section investigates the phenomenon of a distributed identity outside of free jazz, arguing that this phenomenon is omnipresent in music, also in classical European score-driven performance. This is especially evident in the tradition of the string quartet.

– $\Sigma$ –

We should not terminate our discourse on distributed identities without mentioning also some classical traditions, especially the string quartet culture, that are also aware of this gyroscopic effect. It is not only a free jazz topic, but in free jazz it becomes the singular criterium, while for other music cultures it is one out of a number of characteristics.

Our reference to the string quartet theory is [93, Part XIV]. Following the research of Ludwig Finscher, the string quartet is characterized by three components:

1. Four part writing
2. Discourse of four educated humanists
3. Richness of sound qualities of the violin family

For our topic, the second component is important. A string quartet is a dialogue, not only a four-fold monologue. The score alone is not sufficient to convey the message, the two violins, the viola, and the violoncello must interact in an exchange of sounding human messages, but not only abstract score-driven statements. These messages must be communicated as embodied gestures in order to fulfill the composition's identity. Such a gestural interaction is an added value to the score's logic, it is this distributed identity which creates a successful string quartet performance.

**Impact.** The idea of a distributed identity in its passionate reification is central to our understanding of musical time construction. It is an embodied architecture that results from the interaction of human agents. It is indispensable for the creation of a shared reality that is not given by linear imported time-lines.

– $\Theta$ –

# Part IV

# New Developments on Musical Time Concepts

# 11

# Limits of Gestural Diagrams

**Summary.** This part adds the authors' new mathematically conceived musical time concepts and their application to a modeling of Kramer's ideas.

– $\Sigma$ –

## 11.1 The Role of Limits

**Summary.** This section develops the basic mathematical theory of projective limits that are needed for the gestural model of Kramer's vertical time.

– $\Sigma$ –

### 11.1.1 Representable Functors and Limits

In mathematical category theory, one considers co- and contravariant functors of very different types, stemming from mathematically interesting constructions. Such functors are not conceived as being related to particular spaces, i.e., objects, in the given categories. This means that such functors are only defined by a specific behavior and not by the points of a space that would "materialize" such a behavior. For example, if $A, B$ are two objects in a category $\mathcal{C}$, one may consider the contravariant functor $C(A, B) : X \mapsto X@A \sqcup X@B$ mapping object $X$ to the coproduct (disjoint union) of the morphism sets $X@A$ and $X@B$. This is the set of $X$-valued points in $A$ or in $B$. But in general this is not the set of $X$-valued points in a determined space in $\mathcal{C}$. The same construction would be possible when defining $L(A, B) : X \mapsto X@A \times X@B$, the mapping of $X$ to the cartesian product of the two sets $X@A$ and $X@B$, and here again, this set is not, in general associated with the set of $X$-valued points in a space of our category.

A special case of such functors is given by the definition $X \mapsto X@Z$, the set of $X$-valued points in a space $Z$. Such a functor, denoted by $@Z$, is



G. Mazzola et al., *Making Musical Time*, Computational Music Science,
https://doi.org/10.1007/978-3-030-85629-8_11

said to be *represented* by space $Z$. A functor which is isomorphic to a represented functor is called *representable.* In our above examples, if they were representable, we would have representing spaces $C(A,B), L(A,B)$ such that $X@C(A,B) \xrightarrow{\sim} X@A \sqcup X@B$ and $X@L(A,B) \xrightarrow{\sim} X@A \times X@B$. In the category **Ens** of sets, such representing sets do exist, namely $C(A,B) = A \sqcup B$, the disjoint union of sets, and $L(A,B) = A \times B$, the cartesian product of sets.

In our context, we are interested in the following type of functors. Let us first discuss the situation for sets. A diagram $\mathcal{D}$ of sets is a map $\mathcal{D} : \Delta \to \mathbf{Ens}$ of a directed graph $\Delta$ into the category **Ens** that sends every vertex $v$ of $\Delta$ to a set $\mathcal{D}(v)$, and every arrow $a : v \to w$ to a set function $\mathcal{D}(a) : \mathcal{D}(v) \to \mathcal{D}(w)$. The (projective) limit of this diagram is the subset of the cartesian product $\prod_v \mathcal{D}(v)$ of the diagram's vertex spaces such that for any two elements $x \in \mathcal{D}(v), y \in \mathcal{D}(w)$, we have $\mathcal{D}(a)(x) = y$ for any function that is associated with an arrow $a : v \to w$. This limit is denoted by $\mathbf{Limit}\mathcal{D}$. We have all projections $p_v : \mathbf{Limit}\mathcal{D} \to \mathcal{D}(v)$, and the commutative triangles

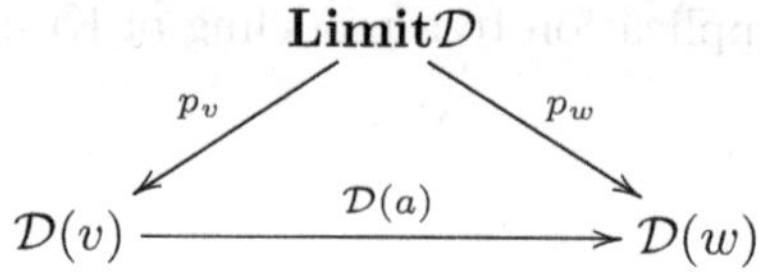


for all arrows $a : v \to w$. The universal property of this limit is that in the category **Ens**, we have a functor $X \mapsto \mathbf{Limit}(\mathcal{D}(X))$, where $\mathcal{D}(X)$ is the diagram of function sets $X@\mathcal{D}(v)$, $v$ the vertices of $\mathcal{D}$, and the canonical transition maps $\mathcal{D}(X)(a) : X@\mathcal{D}(v) \to X@\mathcal{D}(w)$ given by the original maps $\mathcal{D}(a)$. We then have an isomorphism of functors

$$\mathbf{Limit}(\mathcal{D}(X)) \xrightarrow{\sim} X@\mathbf{Limit}\mathcal{D},$$

i.e., $X \mapsto \mathbf{Limit}(\mathcal{D}(X))$ is represented by the limit set $\mathbf{Limit}\mathcal{D}$.

A simple example for sets is the diagram $\mathcal{D}$ defined by a single set map $f : A \to B$, representing the arrow digraph $0 \to 1$. The limit of this diagram is the classical graph of $f$, it consists of all pairs $(u,v) \in A \times B$ such that $f(u) = v$. The general meaning of limits is to represent collections of spaces that are connected by characteristic transition maps.

The situation for limits it a more general category $\mathcal{C}$ is the following: We are again given a diagram $\mathcal{D} : \Delta \to \mathcal{C}$, now into the category $\mathcal{C}$ instead of **Ens**. The transition maps are now morphisms in $\mathcal{C}$. We again have the functor $X \mapsto \mathbf{Limit}(\mathcal{D}(X))$, where $\mathcal{D}(X)$ is the diagram of morphism sets $X@\mathcal{D}(v)$, $v$ the vertices of $\mathcal{D}$. But now, it is not true that this functor is representable by an object of $\mathcal{C}$ in general.

The following section is dedicated to prove that in the category *Gesture* of gestures, limits do exist.

### 11.1.2 Existence of Limits

**Summary.** Here we prove the existence of projective limits of gestural diagrams in the category of gestures.

– $\Sigma$ –

For our theory of a mathematical representation of distributed identity, we work with diagrams of gestures, this was discussed in Chapter 10. In this section, we prove the existence of limits of such diagrams. Recall that a morphism of gestures is a commutative diagram of digraphs

$$\begin{array}{ccc} \Delta & \xrightarrow{\ \delta\ } & \overrightarrow{X} \\ {\scriptstyle t}\downarrow & & \downarrow{\scriptstyle \overrightarrow{f}} \\ \Gamma & \xrightarrow{\ \gamma\ } & \overrightarrow{Y} \end{array}$$

where the digraphs $\overrightarrow{X}, \overrightarrow{Y}$ are the digaphs of continuous functions $I \to X, I \to Y$, the curves on the real unit interval $I = [0,1]$, and $\overrightarrow{f}$ being the digraph morphism induced by the continuous function $f : X \to Y$.

Given a diagram $\mathcal{D} : \Sigma \to$ *Gesture* of gestures, we have the associated limit in the category *Digraph* of digraphs. We should recall here that *Digraph* is a topos and therefore admits limits (even for infinite diagrams). This means that we have a canonical limit morphism

$$\mathbf{Limit}_{\mathcal{D}}\Delta_v \to \mathbf{Limit}_{\mathcal{D}}\overrightarrow{X}_v$$

The codomain digraph $\mathbf{Limit}_{\mathcal{D}}\overrightarrow{X}_v$ is the limit of the digraphs $\overrightarrow{X}_v$, which are essentially the spaces $I@X_v$, so we are looking at the limit of function sets for $I$-valued functions with values in the members of a diagram $\Sigma \to \mathbf{Top}$ of topological spaces $X_v$. But it is known that limits exist in the category $\mathbf{Top}$. This means that we have $\mathbf{Limit}_{\mathcal{D}}\overrightarrow{X}_v \xrightarrow{\sim} \overrightarrow{\mathbf{Limit}(\Sigma \to \mathbf{Top})}$. This means that the limit of $\mathcal{D}$ is given by the digraph morphism

$$\mathbf{Limit}_{\mathcal{D}}\Delta_v \to \overrightarrow{\mathbf{Limit}(\Sigma \to \mathbf{Top})}$$

and this *gesture* is the limit of the given diagram of gestures in the category *Gesture*.

♪ **Example 1** It is important to recognize that a limit can be empty, it is not guaranteed that any point can be exhibited in a gestural limit's space. A simple example results from the limit of a gestural diagram of digraph $\Delta = \bullet \overset{u}{\underset{v}{\rightrightarrows}} \bullet$, where the domain digraph is $\bullet \to \bullet$, while the codomain digraph is again $\Delta$. We take two digraph morphisms, one mapping the arrow of the domain to $u$ and the second mapping that arrow to $v$. This generates an empty limit of these two morphisms,

## 11.2 Temporal Interpretation of Limits

**Summary.** This section develops a temporal interpretation of projective limits of gestural diagrams.

$- \Sigma -$

The theory of gestures admits a slightly different interpretation of a gesture $\gamma : \Delta \to \overrightarrow{X}$ as a continuous function $|\gamma| : |\Delta| \to X$ from the associated topological space $|\Delta|$ to the topological space $X$, see Section 8.3. This adjointness transforms digraph morphisms into continuous functions. In particular, the above limit reads as an adjunction of

$$\mathbf{Limit}_{\mathcal{D}}\Delta_v \to \overrightarrow{\mathbf{Limit}(\Sigma \to \mathbf{Top})}$$

with

$$|\mathbf{Limit}_{\mathcal{D}}\Delta_v| \to \mathbf{Limit}(\Sigma \to \mathbf{Top}),$$

which interprets the limit digraph $\mathbf{Limit}_{\mathcal{D}}\Delta_v$ as a topological space that is associated with a topological colimit of copies of the unit interval $I$, the so-called *skeletal space* of the digraph.

This skeletal space is the limit of the corresponding diagram of skeletal spaces $|\Delta_v|$. In other words, we now have the skeletal space of the limit, which identifies with the limit of skeletal spaces whih are connected by the continuous functions associated with the digraph transition morphisms from the original diagram of digraphs given by $\mathcal{D}$. Figure 11.1 shows points in the limit factors of topological spaces, which define one point of the limit.

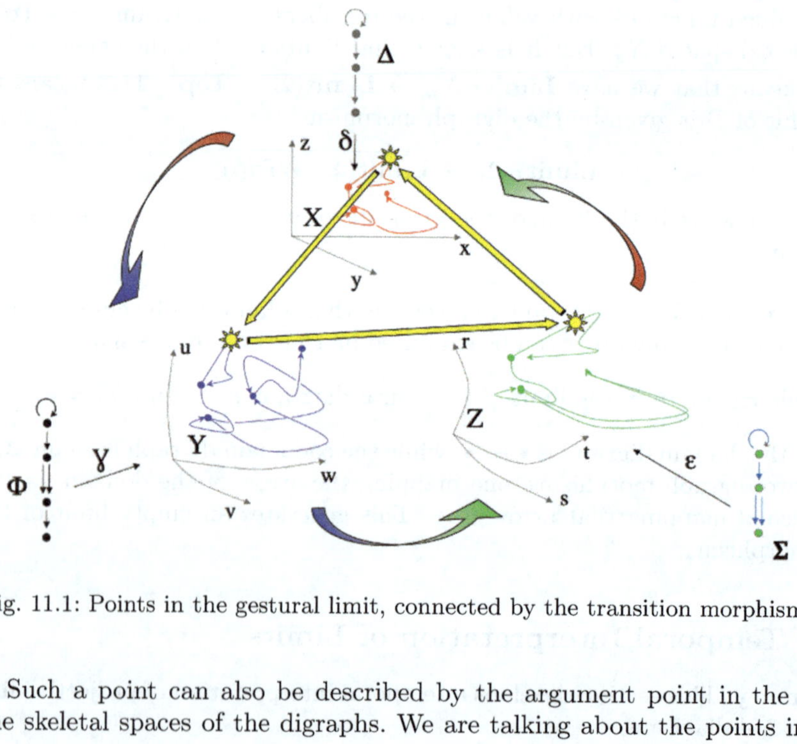


Fig. 11.1: Points in the gestural limit, connected by the transition morphisms.

Such a point can also be described by the argument point in the limit of the skeletal spaces of the digraphs. We are talking about the points in the topological space $|\mathbf{Limit}_{\mathcal{D}}\Delta_v|$. These points are expressing a "harmony" of gestural arguments in this gestural diagrammatic collaboration in terms of our language involving a distributed identity. More generally, we could consider the points $p : |\Delta| \to |\mathbf{Limit}_{\mathcal{D}}\Delta_v|$ valued in a non-empty digraph skeleton $|\Delta|$. The non-automatic existences of such points means the existence of shared skeletal points distributed over the diagram's gestures.

**Definition 1.** *It is this type of points which we now* define *as being the temporal arguments derived from the given gestural collaboration (diagram).*

This definition has two conceptual ressources: the gestural distributed identity descried by the diagram $\mathcal{D}$ and the limit of this diagram. The former is an empirical fact, while the latter is an *a priori* construction from pure mathematics and could be viewed as a Kantian component of our construction.

The next and final step in our conceptualization of Kramer's vertical time and similar non-physical (musical) temporalities will be to consider space-time environments that may comprise the above diagrammatically conceived temporal arguments. But the decisive point here is that time is not given outside of a gestural interaction, it appears as a construction that follows from a gestural collaboration, which also means that *it vanishes after such a collaborative artistic maneuver.*

**Impact.** This chapter develops a precise conceptualization of vertical time in terms of limits of gestural diagrams. This turns the somewhat fuzzy "distributed identity" into a concrete geometric concept, which becomes feasible because projective limits *do* exist in the category of gestures.

– $\Theta$ –

# 12

# Imaginary Time

**Summary.** We discuss the concept of complex time (extending real time by an imaginary component) and its signification for music. The concept of complex time was developed in Mazzola's Topos of Music, vol. III, for a precise description of musical artistic presence.

– Σ –

## 12.1 Complex Time in Music

In Hawking's physical model, time is a complex number $t + i.s$, where $t$ is the traditional time value and $i.s$ is the imaginary component. It is not clear at all whether and how this imaginary component could be a part of human consciousness. Mazzola's emails to Hawking regarding a possible cognitive reality of complex time were not answered by Hawking, and now, it's too late. But we conjecture that our presence of consciousness, where thoughts are built and processed, could happen in that imaginary direction, so that the classical real space-time is complemented by an imaginary time-space defined by imaginary time. If physicists are entitled to introduce new time dimensions, there is no reason to prevent artists from doing the same and claiming that creative human consciousness is hosted in such a time-space that is "orthogonal" to the physical one. This means that at any classical physical time $t$, we would have an entire time-space defined by an imaginary time $i.s$ plus some space coordinates attached to that time.

Based upon such arguments for an extra time dimension, we argue that the concept of presence in time-critical arts requires such an imaginary time and space. It is in this realm where the transitional processing of past music to future music, the planning of gestural strategies, the body-instrument-sound interface are all displayed and organized. It is a quite dramatic change of understanding of what happens in the artistic presence, since it eliminates the mystification of spontaneity and imprevisibility in creative performance. These attributes have



G. Mazzola et al., *Making Musical Time*, Computational Music Science,
https://doi.org/10.1007/978-3-030-85629-8_12

had a great influence on the non-understanding of creative performance, from classical music to free jazz. Improvisation, creative performance—all of that has been boiled down to these negative concepts: Spontaneity and imprevisibility are just negations of any sort of positively defined artistic shaping; they are the "emergent properties" of creativity mysticism, creativity by negation. Telling a musician to be spontaneous is of no help whatsoever: It is just a recommendation to rely on nothing that could be conceived of in the artistic-shaping activity.

## 12.2 Imaginary Time

**Summary.** Imaginary time is a component in complex time in music. It is the dimension, where in our theory of a temporal interpretation of projective limits of gestural diagrams, this time is positioned as an existential utterance.

– $\Sigma$ –

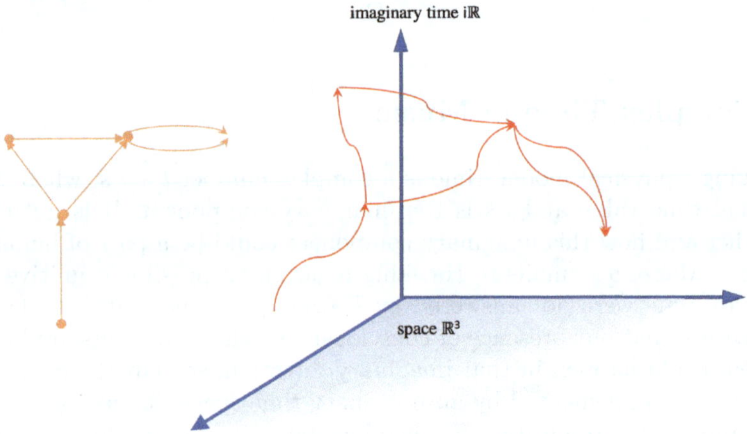


Fig. 12.1: The imaginary spacetime where the temporal arguments of the limit digraph are embedded.

Let us recall the gestural construction of musical time as being realized with the limit of skeletal spaces of diagrams of gestures. The question here is about the realization of such limits in our space-time reality. To begin with, the imaginary time space-time $i\mathbb{R} \oplus \mathbb{R}^3$ is a four-dimensional vector space where imaginary temporal instances can survive. From the discussion in Chapter 11, we learn that representing the gesturally constructed time needs a representation of a limit digraph skeletal space. Such as space is a colimit of line spaces (the real number space $I = [0, 1]$). But to represent such a space, it evidently sufficient to have a three-dimensional real space. Such a space is available in $i\mathbb{R} \oplus \mathbb{R}^3$.

We therefore may realize the temporal parameters in the space $|\mathbf{Limit}_{\mathcal{D}}\Delta_v|$ within the imaginary time-space $i\mathbb{R} \oplus \mathbb{R}^3$. What we are stating is that the temporal space can be realized within $i\mathbb{R} \oplus \mathbb{R}^3$. Moreover, an evaluation of points $p : |\Delta| \to |\mathbf{Limit}_{\mathcal{D}}\Delta_v|$ can be achieved with values in $i\mathbb{R} \oplus \mathbb{R}^3$. The specific nature of such points will reflect the management of temporal data derived from the gestural collaboration in a distributed identity.

And let us also stress that the gestural time construction is by no means the simulation of a linear time concept. The temporal parameters in the limit space can easily be a digraphic combination of line segments. The vertical time in the sense of Benjamin Kramer isn't linear either, it is a mental construction that expresses the complexity of our time deployment, including multiple threads between shared start- and endpoints, or loops.

It is also important to understand the qualitative difference between the gestural parameter, which is essentially the unit interval domain $I = [0, 1] \subset \mathbb{R}$. One could argue that this domain already defines the linear character of time. It is true that the digraph parameter space $I$ bears a linear nature. But this is an abstract ordering of gestural curves. It may be applied to purely spatial gestures without any temporal connotation. This could be what Kant used in his description of time as an "inner a priori sense". But what generates our concept of time stems from the limit of a gestural diagram, not from the gesture as a structure per se. The special case of a singleton diagram of one gesture cannot be taken as a typical example. A single gesture does not constitute a valid temporal entity, it resembles the empty time rather than what our experience of time looks like. One single gesture is the conceptual germ, not the full-fledged time ontology.

**Impact.** This chapter embeds the limit construction of vertical time in a complex time-space $i\mathbb{R} \oplus \mathbb{R}^3$, where the imaginary component relates to the modern concept of complex time as proposed in cosmology.

– $\Theta$ –

# 13

# 3.3. Modeling Kramer's Time Concepts

**Summary.** This section implements our interpretation of the most significant of Kramer's time concepts.

– $\Sigma$ –

## 13.1 Interpreting Gestural Time

**Summary.** This section deals with our interpretation of Kramer's gestural time in our conceptual framework of limits of gestural diagrams.

– $\Sigma$ –

Kramer's gestural time concept essentially overrides the linear and absolute time, replacing time parameters by gestural parameters. Gestures in our setup are driven by parameters different from absolute and linear time, they determine the gestures' processuality, which can be spatial and/or temporal. Linear time would then be an image value addressed by the gestural parameters. It could very well go against the linear time direction, being, for instance, determined by a logical argument of causal or final nature. For example, modulation could be gesturally parametrized from future to past, from the target tonality to the tonality of departure. And the hypergesture of a contrapuntal construction could go vertically from the *cantus firmus* to the *discantus* in pitch direction.

## 13.2 Interpreting Vertical Time

**Summary.** This section deals with our interpretation of Kramer's vertical time in our conceptual framework of limits of gestural diagrams.

– $\Sigma$ –



G. Mazzola et al., *Making Musical Time*, Computational Music Science,
https://doi.org/10.1007/978-3-030-85629-8_13

The gestural time is however not a concept that unites the musical instances to generate an identity, that big picture, which Kramer draws when invoking the integrated perspectives of a sculpture when uniting the movement around the sculpture's presentation.

Keeping Kramer's metaphor, vertical time is an integral of gestural approaches, of components of comprehension. It is more than the sum of those gestural perspectives. Keeping Kramer's metaphor in mind, the integrated information results from the gestural interaction, from the comparison and relationship among all of the sculpture's perspectives.

This concept therefore asks for a precise shaping of such a network of gestures and therefore a construction of vertical *time*. It is more than the gestural architecture; vertical time must be a consequence of the gestural network, not the network alone.

In view of our discussion of diagrams of gestures as related to the construction of a distributed identity in free jazz and other music performance, *we propose to define vertical time as being the topological domain space of the limit of a gestural diagram, and, more generally, as the morphism of a digraph's topological space into that limit domain.* We refer to our detailed discussion of this setup in Chapter 11 and its embodiment in the imaginary space-time in Chapter 12.1.

**Example 2** A first small example of such a vertical time construction from the limit of an elementary 'linear' gestural diagram, Figure 13.1. This limit is the diagonal line in this figure's square to the right. It shows the pairs of corresponding values from lines $c$ and $c'$. This time construction is an elementary 'communication' between the two involved simple lines. Ontologically it is more than the single lines, it describes a correspondence of gestural parameters that generates a temporal *collaborative* quality.

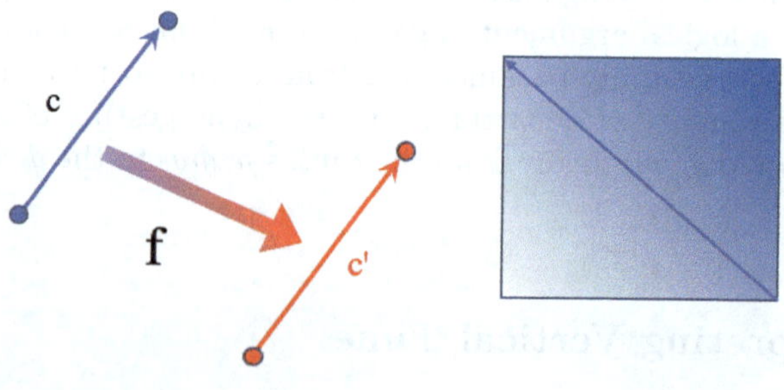


Fig. 13.1: The topological space associated with the limit of the two line digraphs.

**Example 3** A second small example of such a vertical time construction results from the limit of an elementary 'circular' gestural diagram, Figure 13.2. This example shows the limit as being the red line in the torus of this figure. It is essentially the same ontological construction as in the previous example, except that the loops now connect in a torus, cartesian product of two circles, instead of a square as before.

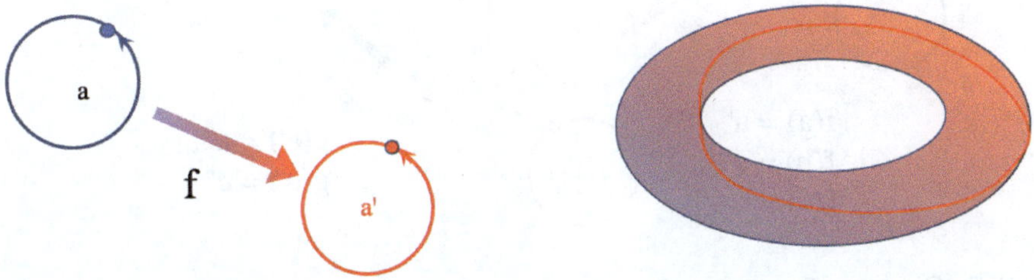


Fig. 13.2: The topological space associated with the limit of the two circle digraphs.

**Example 4** An example of such a vertical time construction from the limit of a simple, but slightly more involved gestural diagram is shown in Figures 13.3 and 13.4.

We start with a rather simple diagram $\mathcal{D}$ of three gestures, which are connected by three morphisms $f, g, h$ as shown in Figure 13.3. The limit of this diagram, when transformed in the equivalent display of a continuous map from the topological space associated with the limit of the three digraphs is shown in Figure 13.4.

This representation needs some comments. First, the limit of the two loops $a, a'$ is shown as a line on the torus, which is the cartesian product of two circles. Then, the image of line $b$ on $a'$ identifies with the line on the tube that is attached to the inner circle of the torus, corresponding to the limit of $b$ and $a'$. To this line $(b \sim a')$ we attach the line $c \sim c'$ corresponding to the limit in the cartesian product of $c$ and $c'$. We also show the identification of these points with points from the third digraph $(c' \sim c", a" \sim a''')$. This configuration of three lines is the limit of our diagram's topological domain. It is the time-line that is generated following the limit construction. As for every digraph, this line can be represented without overlapping parts in $\mathbb{R}^3$.

*Remark 2.* It is difficult to compare our construction of vertical time to Kramer's concept because he does not present a precise conceptual definition (and he admits it). Our main concern is to apply our approach to concrete musical situations and to show that it effectively helps understand human time construction in a deeper way.

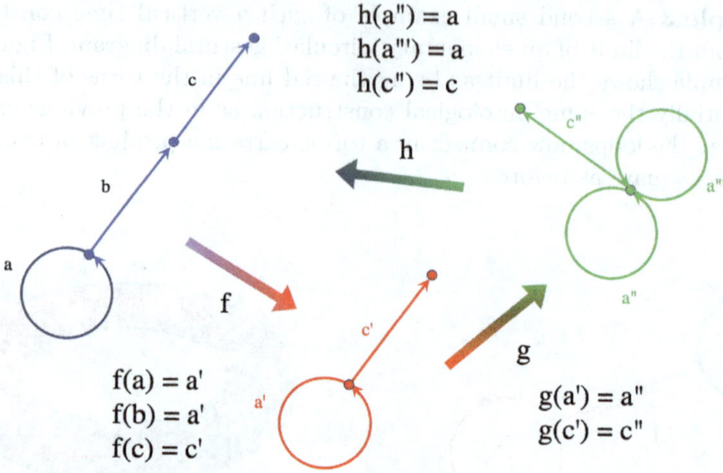


Fig. 13.3: Diagram $\mathcal{D}$ composed of three gestures and three connecting morphisms $f, g, h$, as defined by the images of their arrows.

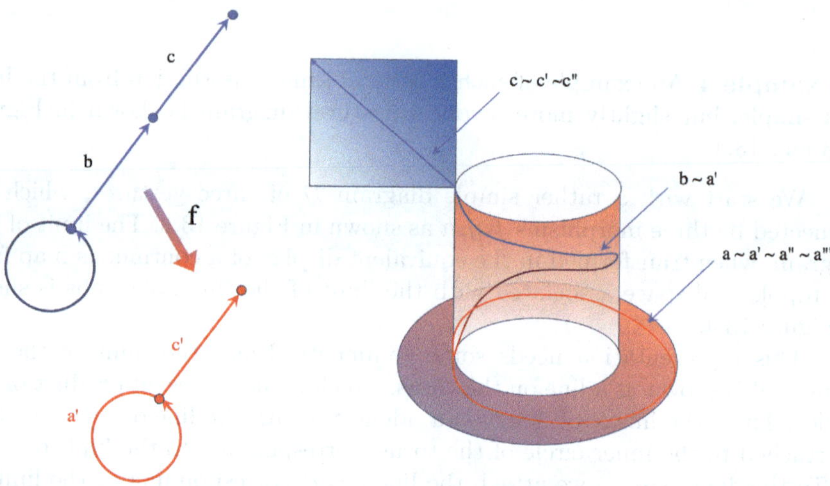


Fig. 13.4: The topological space associated with the limit of the three digraphs of $\mathcal{D}$.

It is however evident that our construction generates a non-linear time reality, and it also takes care of the relational nature of human time construction: a projective limit is the perfect mathematical framework for complex relational configurations.

# 14

# Semiotics of Time

**Summary.** From semiotics it has become clear that gestures can be presemiotic generators of signs. In this chapter we therefore discuss the semiotic and presemiotic position of gestures.

– $\Sigma$ –

## 14.1 Summary of Functorial Semiotics

**Summary.** Semiotics has been modeled in the framework of categories and functors in one of Mazzola's most recent publications, we give a summary thereof.

– $\Sigma$ –

In the framework of a mathematical theory of semiotics, which was introduced in [106], we argued that in the genealogy of semiotics, there is an essential presemiotic mechanism, namely that hinting gesture from expression to content. This was also stressed by Tommaso di Campanella, see also [106, Section 2] and [96, Section 57.3.2].

So gestures seem to play a crucial role in the construction of meaningful entities. In our approach, this basic gestural mechanism was formally shaped in the sign's $S$ essential architecture as shown here:

$$\begin{array}{ccccc} Ex(S) & \dashrightarrow & Sg(S) & \rightsquigarrow & Ct(S) \\ & \searrow & \downarrow & \swarrow & \\ & & S & & \end{array}$$

It shows the sign as a triple diagram: from its expression to its signification 'machine' to its content. This elementary architecture is enriched by the Hjelmslev principles of vertical extensions, namely, that signs may in their three components be signs in their own right, which for the expressive level



G. Mazzola et al., *Making Musical Time*, Computational Music Science,
https://doi.org/10.1007/978-3-030-85629-8_14

is called *connotation*, for the content level it's called *meta-theory*, and in the middle signification level it's called *signification*. This generates what we called Hjelmslev jets, H-jets for short, i.e., sequences of arrows that extend the simple sign diagram up or down. Here is such a H-jet graph:

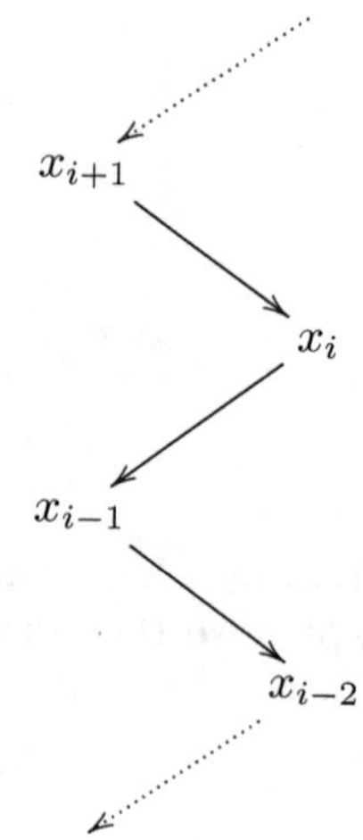


For example, the double articulation in language states that the acoustical content $Ct(X)$ of a graphical word sign $X$ is the expression $Ct(X) = Ex(Y)$ of the logical content $Ct(Y)$.

These H-jets are—except for a number of mathematically motivated refinements which we may neglect here—the *objects* of our semiotic category $\mathcal{LDC}$. The category's *morphisms* are paths of arrows of a path category, which are essentially those arrows $(x.) \to (y.)$ between H-jets $(x.), (y.)$, which are defined by the two arrows in the elementary sign architecture:

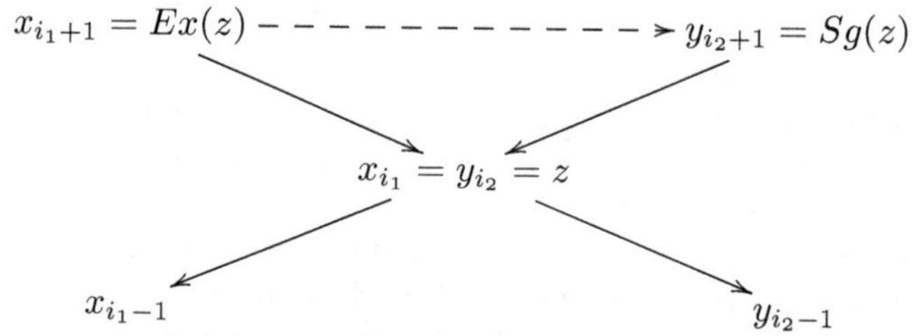


or $x_{i_1+1} = Sg(z)$ and $y_{i_2+1} = Ct(z)$:

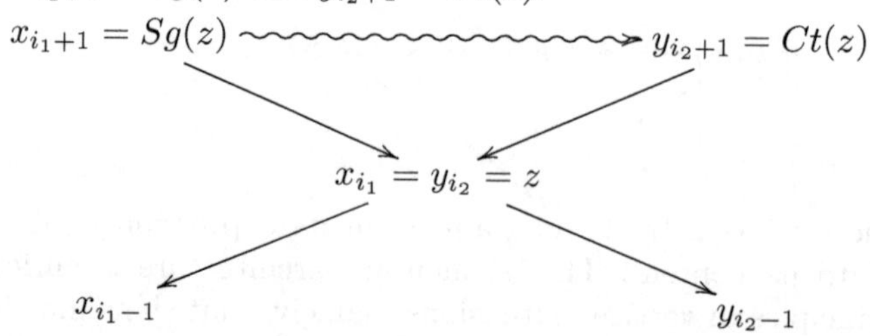

## 14.2 Gestures in Semiotics

**Summary.** This section exhibits and discusses the crucial role of gestures in (the functorial theory of) semiotics.

– $\Sigma$ –

In this functorial semiotics, gestures should be dealt with a special attention, in fact, gesture theory distinguishes two types of gestures: wild and tamed ones. The tamed gestures are those, which function as expressions of semiotic contents. Many socially present gestures are tamed, for example the O.K. gesture of the hand, closing the thumb and the index to a closed circle. Linguistically oriented US-American gesture theory, such as developed by Adam Kendon and David McNeill, [76], focuses on tamed gestures. But the French approach to gesture theory rather focuses on wild gestures, which may be generators of semiotic contents, but are not conceived as their carriers, see our discussion of the French tradition in [96, Chapter 58].

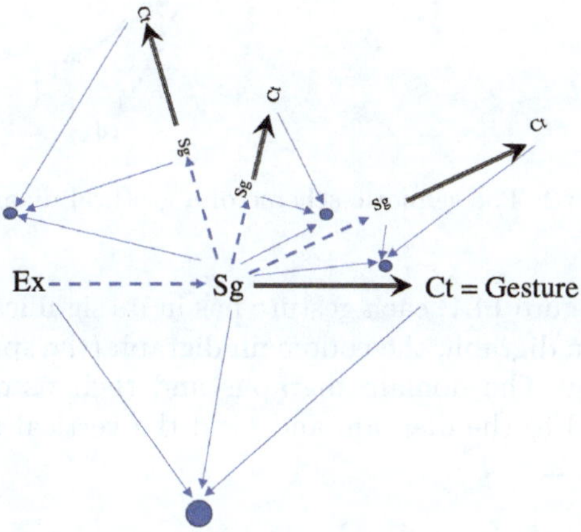


Fig. 14.1: A gesture as a content that has its signification level as being a homonymous expression of three different contents.

What does it mean that a gesture is not an expression of another entity? It means that there is no direct morphism from the gesture's H-jet to another H-jet, i.e., a morphism that puts a gesture to the expressive position within a H-jet. This does not prevent the possibility that if a gesture is a point within a H-jet, it may give rise to a morphism higher or lower in the H-jet's array. This means that, for example, a gesture might be the connotative component of a H-jet point, which may be the expressive part within its own context. Figure 14.1 shows such a complex combination of a gesture as a content that has its signification level as being a homonymous expression of three different contents.

In the context of our gesturally constructed musical time, we first need to set up the semiotic skeleton of a gestural diagram. This works as follows: the diagram's expression $Ex(\mathcal{D})$ points to all morphisms $\delta_i$ between gestures, and these morphisms point to their domain and codomain gestures $d_i$, $cd_i$, and the associated map graphs $\Gamma_i$, this is the usual definition of a function: domain, codomain, and graph, see Figure 14.2

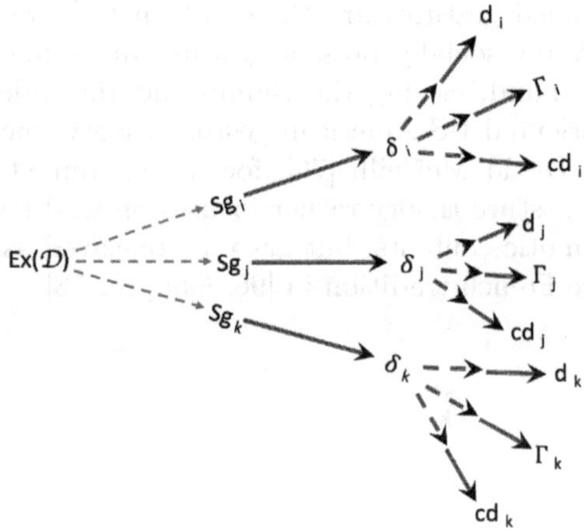


Fig. 14.2: The semiotic scheme of a gestural diagram.

As shown in Figure 14.1, each gesture has in its signification point a triple arrow: to the domain digraph, the codomain digraph (the spatial digraph), and the function's graph. The domain digraphs and their associated topological spaces are connected to the diagram and yield the vertical time structure discussed in Section 11.2.

## 14.3 Time Gestures: From Gestures to Signs

**Summary.** This section discusses the semiotics of time constructions within the gestural theory of time.

– $\Sigma$ –

The above presentation of the semiotic architecture of the gestural diagram shows that the gestures are not understood as being expressions of contents, they are only content components of morphisms, but not expressions, and the diagrams of digraphs are not taken from gestures, but from their signification conditions.

This means that the gestural construction of vertical time does not 'tame' our gestures, it only connects their signifying digraphs' components. In our setup, gestures remain wild animals, and their temporal output is taken from their signifying digraphs, not from the gestures as signs. Gestures here remain contents of signs, not expression of signs, they don't signify.

## 14.4 Time in Creative Semiotics

**Summary.** This section discusses the dynamical nature of semiotics in terms of time.

– $\Sigma$ –

Functorial semiotics is a category-theoretical approach to the theory of signs. As such, time does not intervene as an explicit variable. In category theory, time is absent insofar as categories are abstract entities in a mathematical "Platonic sky". However, the nature of semiotics is a dynamical one, see also our discussion of this aspect in [106]. This means that the universe of signs is not fixed once for all. The procedure of creativity is by definition the action where new signs are added to the given repertory in the category $\mathcal{C}$ of signs.

This creative action defines a very abstract time dimension, defining a succession of the category's states, $\mathcal{C}_1 \subset \mathcal{C}_2 \subset \ldots \mathcal{C}_n$ It is not a musical time concept here, it is a logical and discrete time, a succession of states of a category that has nothing to do with what in music is constructed.

# 15

# Goebel, Pang, and Rochester: Applications

**Summary.** This chapter presents three applications of the time theory to compositional creativity.

– $\Sigma$ –

## 15.1 Jordon Goebel: Exploring the Construction of Time through Music Composition and Compositional Techniques

**Summary.** Applying prior explained time concepts to a compositional process to invoke multiple layers of time awareness, resulting in a finished piece of original music. Altered notation will be experimented with to better explain and achieve these temporal layers. Examining ideas of this composition as an experiment for a performance practice on similar new pieces, while hopefully inspiring ideas of performance practice on old repertoire as well.

– $\Sigma$ –

### 15.1.1 Introduction

Music constructs and manipulates our immediate experience with time and perception of time. These impositions are subjective and can be positive or negative experiences for the listener; however, effectiveness is crucial in order to engage the listener. One tends to associate "Good" or "Effective" music with music that can achieve some level of engagement. Goebel believes that our experience with time is one that is based off of the duality of what people construct as physical time or as (Johnathon Kramer calls "Linear time") and how it simultaneously interacts with each person's personal perception of that time. Goebel views these "times" as an objective versus subjective and this a direct result of the constructs of modern Western society. From this a dissonance is



G. Mazzola et al., *Making Musical Time*, Computational Music Science,
https://doi.org/10.1007/978-3-030-85629-8_15

created in the human soul, one that people are constantly looking to patch through means of media, activity, work, love, etc., in order to escape these self-created confines of linear time. This has fascinated Goebel as a composer and musician, and he began to fixate on this existential experience of transcending time. He also concludes that it seems as if most people acknowledge a varying perception of time through various activities, but the thought ends superficially. Therefore, this composition project was created in hopes to go further than an acknowledgement of this phenomena. Instead direct focus to invoke this experience, in hopes to create a more rewarding and intimate relationship with the music we engage with.

In 2019, Goebel began the process of composing a solo piano piece that was to dive directly into the thoughts and concepts that are discussed in this book, regarding perception and temporality. Through this experience, it also became apparent that there was more to be explored than just a cut and dry idea of success or failure as suggested above. However, it is important to start at this foundation. It is also worth stressing that by no means is this experiment looking for direct empirical evidence in this theory, but instead, wanted to apply these concepts. The goal of this study is to compare and contrast approaches in how to have these topics influence the way performers and composers move forward with their experience and creation of meaningful music, not just as producers of music but also as consumers of the product and the journey. Therefore, time perception and manipulation should be present for the listener of the piece, but also for the performer of the piece as a performer and listener simultaneously. This is a similar take to the philosopher Merleau-Ponty, regarding perception being influenced by the experience and in this case the experience of the engaged listening or the experience of partaking in the creation of the music. Due to this, there is a unique duality for the performer as their roles change from consumer/listener to performer/creator throughout their interactions with this piece.

### 15.1.2 Explanation of the Piece

Perspectives is a solo piano piece that is meant to challenge the performer into exploring two dimensions of time simultaneously. One level of time being the physical or linear time that is written on the page as in most Western music. From left to right, there is a basic linear direction of time from start to finish and the tempo at which is occurs at. This is notated in a typical linear grand staff that is to move in a completely expected way; durations, meter, and tempo are all to be present. This grand staff in the piece is labeled "LP" for linear perspective of time (see Figure 15.1 lower grand staff). To contrast this, a second grand staff was created to be played by the same performer, but with different temporal properties. Goebel titled this grand staff as "PP" or personal perspective of time (see Figure 15.1 upper grand staff). This staff is to be interpreted by the performer, pitches are given, and approximate durations are present, however there is no given meter, bar lines, phrase markings, or articulations. The personal perspective should come from within the artist and not be heavily dictated by the composer or the sheet music, but instead a guide. However, in order to invoke this duality, it is necessary to have these ideas become materialized and put against this level of linear time in order to encourage the inter dialogue of the performer to invoke their own level of transcendence.

Structurally the piece is constructed of 12 moments, and as these moments are recreated, they should have a life that is dependent on the performer who is realizing them which is due to the ambiguousness of the personal perspective grand staff. This is also crucial in the layout of the notation in this piece each moment is directed in different ways in order to interact, isolate, or respond to this duality of time. The melodic and harmonic content of the moments are organized and are directed, however are not the focus of the piece. In large scale the piece starts and ends with familiar pitch collections and as the moments progress in a series of fifths, this begins to close in on itself, creating dense clusters in the middle of the piece. This is also the case for flow and tempo, as the beginning of the piece is free and light and develops into a faster tempo and pace that allows for new interactions from the personal perspective. In order to more deeply explain the time interactions of the piece, all twelve of the moments will be broken down.

### 15.1.3 Twelve Moments

The *first* moment in this piece starts solely in the personal perspective. Challenging the performer to start a process of slight improvisation that was to invoke a sense of deep thought and reflection of what is going to happen with this piece music. Through this process Goebel hopes to immediately persuade a personal connection to the musical product that is being created, as well as promote a new musical experience with every performance of the piece. As this the piece begins, it was crucial to start from silence and build the sonic environment. This builds into a bouncing melodic space that is trying to go somewhere, yet always seems to fall back. Initially, Goebel wanted this to invoke that inner personal realm, one that he tends to, "just barely grasp but always falls back to reality."

As this moves to moment *two*, it does just this and fully pulls the performer back to reality with the introduction of the first linear perspective interjection. As this moment starts fully linear, it is then met with small injections of the personal perspective, reminiscent of this free space in moment one. This is also included in performance notes (see example on Figure 15.3) entitled "transitional" as this just a small interjection of the two perspectives colliding.

Moment *three* is included with the performance note, "Relapse" as this linear space has been taken away and now, brought back. Goebel chose "Relapse" while considering its connotations. This word tends to carry a negative connotation that includes feeling of failure or recurring. This is important at the beginning of this piece, as it illuminates the cultural norm of focus and the importance of productivity and this as an inner struggle. This works side by side with the personal and linear time and serves as an arcing structure of the piece. Moment three is almost fully in the personal perspective, other than the metered long tones in the left hand. This shows the present linear reality and the awareness to this reality, even in an almost fully free moment.

In moment *four*, this awareness and lack of productivity brings a level of discomfort or anxiousness. Included in moment four is the performance note "Stirring", with hopes to provoke a sense of anxiety, as these dualities of time perception are now not only experienced but acknowledged. Goebel finds this to be a common experience for this subject, as one becomes aware of the different feelings and perceptions of time through their life, it can be uncomfortable as there are no perfect answers of why this occurs. Just as he begin to wrestle with these concepts and emotions, Goebel is thrown right back into the rat race of life.

Moments *five through seven* are all lead by a metronomic pulse. This was intended to resemble a Monday through Friday work week, as the time passes, there are only a few interjections of thoughts. These thoughts were mimicked in this section with a melodic personal perspective.

This experience is numbing and, in a way, pushes time faster until it lifts again in moment *eight*. This section was meant to be a release of this linear dominated space. This release, when followed by a strong tension as moments five through seven create, induces an almost romantic experience. Time seems to slow down and hang on to every fragment of consciousness.

In moment *nine* through eleven, reflection is in order. Multiple notes are repeated in the personally perspective and meant to revisit the thoughts of this experience. There are motives that are developed and reintroduced however they are all in the personal perspective. In this space the performer can also reflect and add their own thoughts to the piece in an almost entirely free space. As this goes on the music comes back to a developed version of moment one, where the piece meets its full circle in many senses and closes as it began. This structural narrative as described was constructed in hopes to highlight Goebel's personal experience with time duality on a day to day basis, while simultaneously invoking this same time duality through music.

### 15.1.4 Case Study

In order to find results in the success of this piece, it is necessary to put together an idea for a hypothetical case study for this piece to be completed later. To start, it is important to first acknowledge the determinants of success, biases, and reflection prior to these experiments. Success would be determined by having the performer feel a relationship with time in this piece that is personal. As the performer embarks on the journey with the piece, there should be a level of growth that should occur with repetition of this piece and should never end. In an even more ambitious and broad sense, as their personal constructs of life grow, so will their interaction with the piece. This was a focus of the first half of this book, and it is important to acknowledge the incredible number of variables that must be considered when breaking down time and perception. Instead of trying to eliminate variables in the experiment to get a more empirical representation of data, it is more important in this experiment not to remove variables but instead acknowledge them. The addition of a level of improvisation to the "PP" staff (or personal perception) will highlight these individual performer differences, whatever they may be, and hopefully promote that relationship.

Goebel also wants to acknowledge the explanation that is given to the performers prior to the start of their experience with the piece, including even as much as the performance notes may be a hindering bias to the success of this experiment. He wants to be extremely vigilant of the bias that he may impose on the performers by saying too much. However, on the opposite side, too little of an explanation of the concepts and ideas and the performers may never fully engage due to lack of clarity. This may be the toughest and most sensitive predicament of the experiment. It is also important to acknowledge the stage in the performer's career with music.

It is expected that there will be a different experience with pianists of different academic backgrounds. It may be possible that the level of influence that happens in the academic world may be a barrier to overcome in the engagement of improvisation and creative control that is present in this piece.

There are several conclusions that can be taken away from the performance of this piece and that should be held as the focal point of the experiment. First, it is important to note that the overall "quality" of the piece will influence the engagement with the concepts. If the piece does not engage in a temporal experience that is apparent, then the concepts are just added philosophy as a precursor to a "boring" piece of music. As this may seem harsh, it is crucial to the achievement of the goals in this experiment. *If the piece does not invoke a creation of time, then how can a performer be expected to interact with the creation?*

It is important to find performers who are willing to truly engage with the concept. I think these ideas can only be successful through the embodiment of personal experience, again much like the ideas from Merleau-Ponty. If the goal is to invoke a musical experience that is to strengthen a personal connection to a performer's relationship with time, then that performer must be willing and wanting of that experience. This is important as well in the musical success of the piece but should be strictly considered after and not before. Engagement will result in a higher quality of musical content, but this engagement cannot be forced and needs to be emphasized.

### 15.1.5 Layout of the experiment and emphasize on the importance of various performance notes and the significance

Due to the conclusions listed above, the focus of this experiment will be on two different main factors on a group of four performers broken into two groups of two. First is the way the material is introduced via performance notes. All four of the performers will be introduced to the piece the same way; full score with instructing performance notes that will be listed below. For any initial thoughts or concerns they may have, they will be asked to privately log during the first week. At the end of the first week, a performance will be recorded from each of the performers and a series of questions will be followed the performance. These questions will consist of:

1. What is your temporal interaction with this piece?
2. Do you feel as if you are in control of the creation of "time" in this piece?
3. Do you sense a duality of linear and personal time within the different Grandstaffs in this piece or does this just feel confusing through the notation?
4. What are your conclusions from working with this piece thus far, do any of these reflections provoke any thoughts on the way you construct time through musical performance, if so explain.

After this is answered through written documentation, all of the performers will be able to record any questions or thoughts further than what these questions address.

For two of the performers, Goebel will answer these questions to the best of his ability and for two of the performers he will not. This model will be continued for a total of four weeks resulting in four recordings from all the performers. Again, he will ask all of the previous questions, however, as the weeks go by he will also ask the following question:
5. What was your take back or experience from engaging in this piece again for another week; did you make any personal changes to the music if so why?

It is then important to cross reference each of these recordings for differences. There may be differences that the performers notice which will be reflected through their records and there may also be differences that the performers do not immediately recognize. These may include parameters such as total length of the piece, tempo or speed of certain moments, and so on.

At the end of the month of engagement with the piece, a conclusive interview will be held where discussion can be had with each of the performers. The group that was not allowed to have their questions answered will be asked if in post, they appreciated this self-exploration or if instead it was frustrating and why. I will also inform the other group of this variation in the experiment and invoke questions of why they might have preferred one experience over the other and if they feel as if it hinders or promotes their personal relationship with the construction of personal time within the piece.

When finding the performers for the experiment I will first ask a basic questionnaire to try to gather more information on the predetermined tendencies of each performer based on their answers.

1. 1. When you perform a piece from Classical era repertoire, what are the references you look for in the exploration of this performance?

2. 2. What kind of music do you specialize in and examples of repertoire you have performed.
3. How was your academic experience and the way that your instructor interacted with your musicality?

### 15.1.6 Score Reference and Performance Notes

Goebel hopes through this information he can create a base for the application of the concepts that have been discussed in this book. These questions and records will not be concrete evidence on the interactions of the time duality, but instead a look at four different pianists and the way they approach these concepts through this specific piece of music. This hopefully should strike a dialogue in the way that this can be explored and open a door to even more in-depth future experiments.

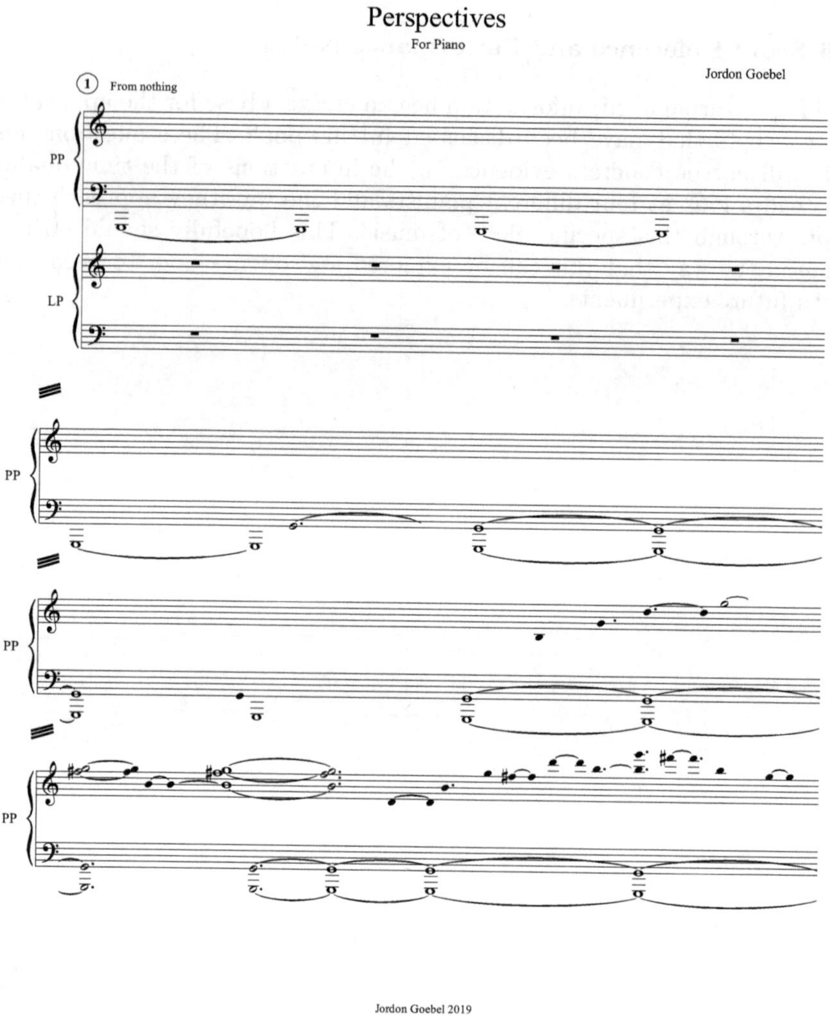


Fig. 15.1: *Perspectives* p.1., ©Jordon Goebel

Fig. 15.2: *Perspectives* p.2., ©Jordon Goebel

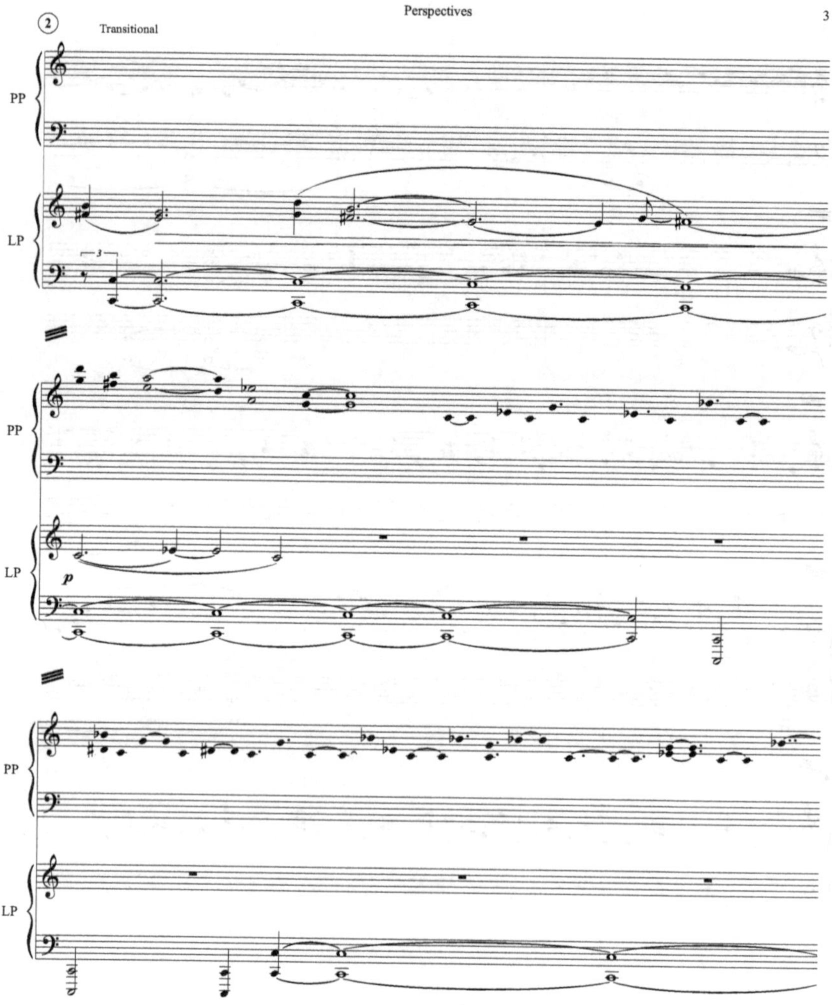


Fig. 15.3: *Perspectives* p.3., ©Jordon Goebel

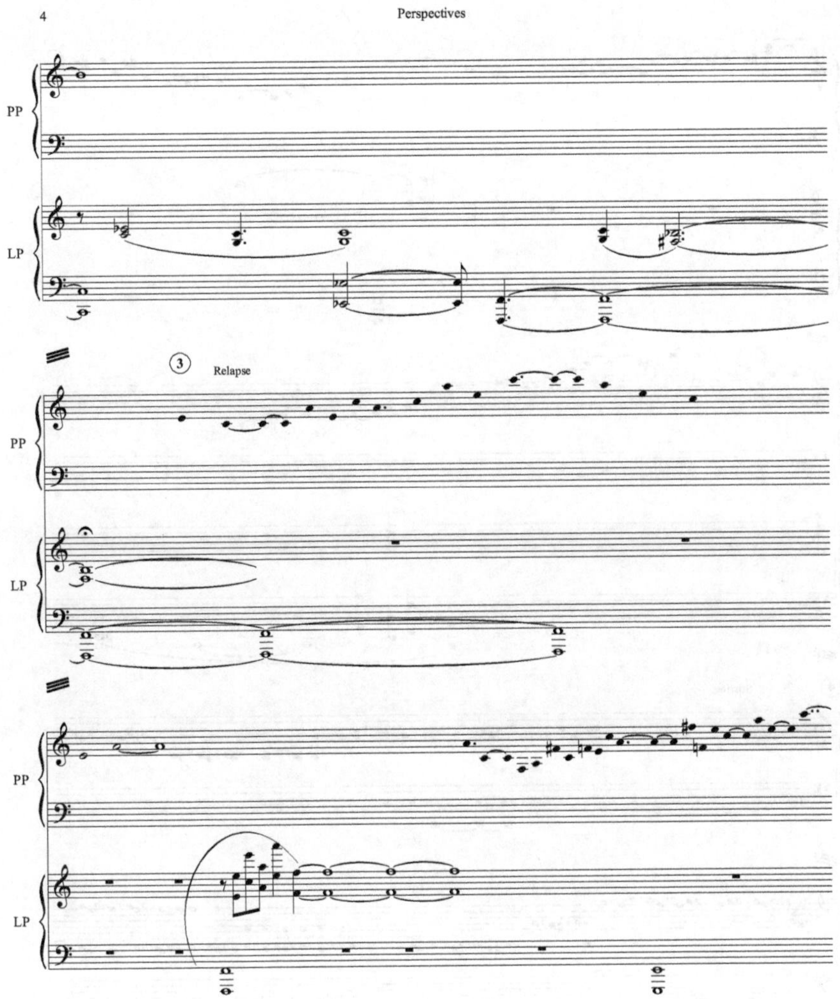


Fig. 15.4: *Perspectives* p.4., ©Jordon Goebel

Fig. 15.5: *Perspectives* p.5., ©Jordon Goebel

Fig. 15.6: *Perspectives* p.6., ©Jordon Goebel

Fig. 15.7: *Perspectives* p.7., ©Jordon Goebel

Fig. 15.8: *Perspectives* p.8., ©Jordon Goebel

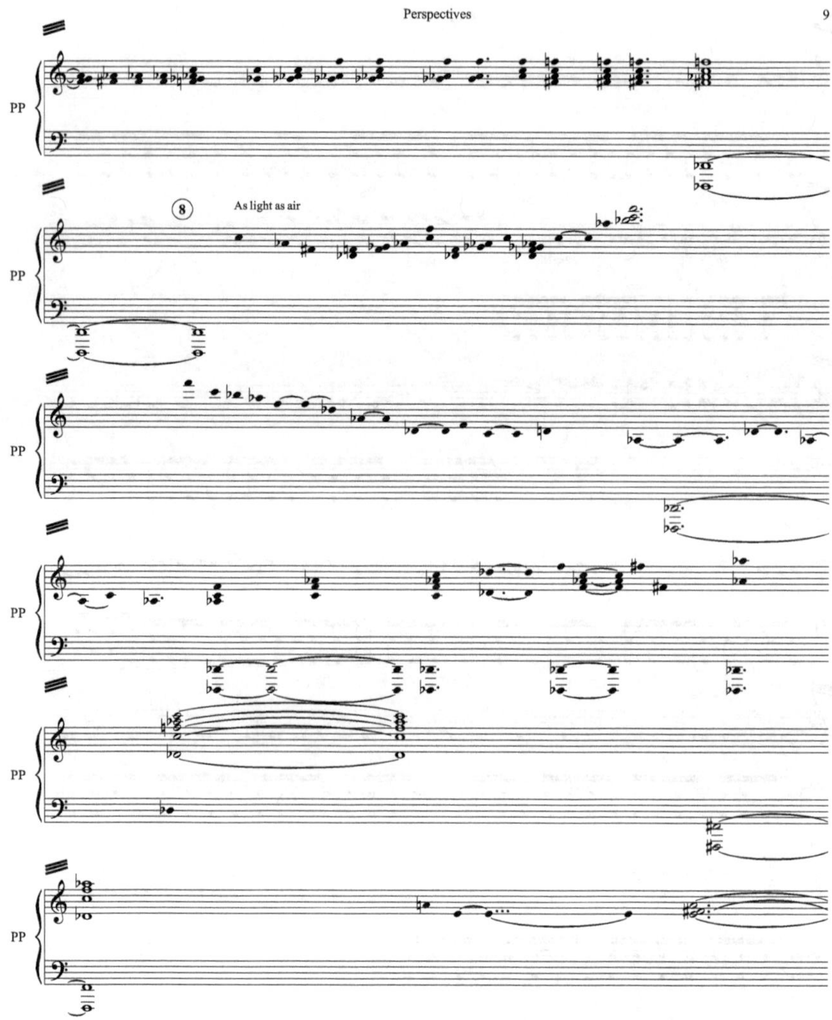


Fig. 15.9: *Perspectives* p.9., ©Jordon Goebel

Fig. 15.10: *Perspectives* p.10., ©Jordon Goebel

Fig. 15.11: *Perspectives* p.11., ©Jordon Goebel

Fig. 15.12: *Perspectives* p.12., ©Jordon Goebel

Fig. 15.13: *Perspectives* p.13., ©Jordon Goebel

## 15.2 Yan Pang's Application/Composition–*Dancing Upstream*

**Summary.** *Dancing Upstream* is a composition by Yan Pang that follows prior explained time concepts (Chapter 4.4). This composition is a product of Pang's ethnographic research study in Sichuan, China, see also [143]. The *Dancing Upstream* music production combines Chinese traditional music with American contemporary dance. By examining this new composition for performance practice, Pang hopes to inspire ideas of performance practice. This chapter's reference, *Dancing Upstream*, focuses on the production process and how the time is conceived in music production.

– $\Sigma$ –

The gestures of dance influence the music production process. Traditionally, music is created, and then dance is created in a followup action or vice versa; however, in *Dancing Upstream*, the order is flexible, which means, sometimes the dance gestures are created and then the score. According to composer/improviser Mazzola, the "making of art" involves the unfolding of mathematical concepts in physical space and time, allowing musical communications and exchanges between performers via "musical gestures" [96]. The theoretical approach to gestures started by Mazzola finds countless applications in the creative sense. One way of theorizing music through gestures is rethinking the role of the score. While writing a musical score, we need to keep in mind that music can also exist without such a structure.

A score is a notational tool that helps performers remember sequences of points to hit while moving continuously in space and time: pick specific pitches at a particular time with that loudness with that timbre. However, the movements' instructions hidden in a musical score need to be transformed into real movements in real-time. It means that performers while playing, "unfreeze" the "frozen gestures" suggested by the score [96]. The musicians needs to understand how to make music and musical gestures without scores in order to better comprehend the meaning and the importance of these two elements when accompanied by a score.

The composer's process sometimes involves a transformation of physical reality (dance) to symbolic reality (score) which is important to the music production process. The composer can start from real physical movement—such as piano improvisation—or 'virtual' physical gestures imagined to be the final result when the music is performed. Starting from physical gestures, the composer has to find the optimal symbolic approximation of their ideas. It means they will answer the question: what are the symbolic indications whose physical realization is closer to the composer's idea?

In this project, Pang relied on this collaborative and recursive process both in performance and composition. On the music-performance side, Pang employed the pentatonic scale and drew on the Chinese rhythmic structure. Quantifying the fluid rhythm to Western classical notation on the score while keeping the "free" flow (see Figure 15.16). For example, in traditional Chinese music, it sounds like there is no rhythm, just musical elements flowing around. It is not measured, instead, it is more like an irregular pulse, which is rhythm, but with a different time-sense than Western classical compositions. The performance demonstrated the result of dancers and musicians negotiating the relationship between gesture and musical expression.

Yan Pang studies how people comply with the rules and break boundaries and investigate the practice of remaining flexible as a mode of inquiry. She uses these aforementioned strategies to inform her practice of improvisation in music and movement to add a unique touch, see Figure 15.14.

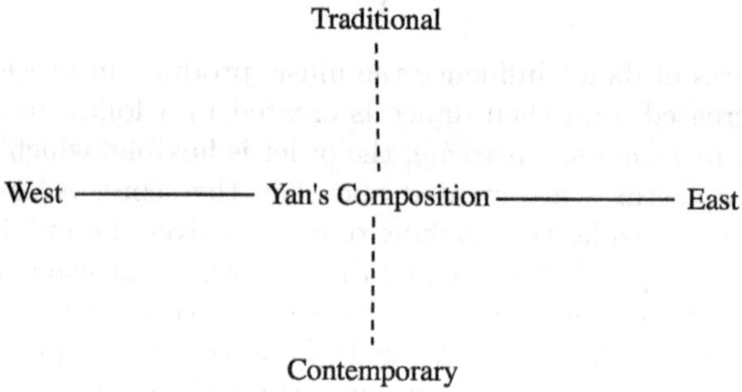


Fig. 15.14: Pang examines the relationship and combination between traditional Chinese folk melodies and modern Western classical compositional techniques, including harmonizing and structuring music elements, which results in a hybrid cultural and musical exchange.

Collaboration enriches traditional music because it centers on different time modalities. It all starts with the events that happened around us having their own meaning. People can draw from these meanings in their own way, producing beautiful forms, full of the body of time. As previously mentioned Mazzola's collaborative work, tension and release create a dynamic and recursive composition through time, like a fractal structure.

Western classical music mainly relies on scores to enable the transmission of musical knowledge across time and groups of people. Composers are expected to make written works more efficient and performances more precise. Even folkloristic pieces throughout the world have often been transcribed into scores to help the work of scholars and performers. However, there are many other musical contexts where the transmission of the repertory is mainly oral, and the production of music itself does not require written scores.

Pang applies these ideas from Mazzola's theory to her own work, in particular *Dancing Upstream* which is a product of her fieldwork in China. This project was conducted in a diverse range of settings that are semiotically constructed realities in a specific time and space. How the field-to-media team co-occupied and constructed such places and moments was partly mediated by technology, technique, and tradition. For example, in order to complete a quality video production, people must take on fairly well-defined roles in relation to each other, much as is the case in a family, village, political institution, workplace, band, and so on. In most of Pang's *Dancing Upstream* project, collaborators were involved in the entire range of roles: producers, directors, composers, production designers, scriptwriters, artists, musicians, on-screen talent, editors, audio designers, audio engineers, and so forth.

Yet, it is often assumed that colleagues in the field hold *a priori* cultural reality and therefore have a set of needs, problems, and interests to which we can and should respond as ethnographers. The traditional ethnographic goal is to capture collaborators' perspectives while being reflexively aware of how one's own orientations diverge from those world-views and thus influence our ethnographic representations of studied communities. The *Dancing Upstream* project takes a decolonizing approach that includes the ethnographer as collaborator and not the all knowing leader. Since the 1960s various ethnographers have noted that we do not study pre-existing, static realities or 'cultures' in the bounded or synchronic sense. Culture is not a thing to be identified, but rather an active semiotic process.

Therefore, creative ethnography is a creative process as well, taking place in a liminal, fluid, and polysemic space. Referring to that space as 'the field' may retain too much of the positivistic, natural history origins of ethnography. 'A moment of time and place' may be more accurate than the more static implications of 'the field'. In telling stories from 'the field', we relate Rashomon-like[1] Rashomon's story, every element is largely identical, from the murdered samurai speaking through a Shinto psychic, the bandit in the forest, the monk, the rape of the wife, and the dishonest retelling of the events in which everyone shows their ideal self by lying. The film is known for a plot device that involves various characters providing subjective, alternative, self-serving, and contradictory versions of the same incident. Rashomon was the first Japanese film to receive a significant international reception. The Rashomon effect is named after the film. In studying culture, we are examining constellations of imagined collectivities [124], mostly in the form of stories.

In *Dancing Upstream*, Pang references Sichuan opera style and uses Western classical atonal techniques. In recent times, many modern music fans have forgotten about the Sichuan Opera and few contemporary operatic compositions are written in the style of traditional Sichuan opera. In addition, not many people outside of China know about the Sichuan Opera that has a long history. Sichuan Opera teaches the listener not only about music, but also about Chinese art, society, history, and people. Although Sichuan Opera is not easily referenced, Pang uses elements from it to bring the legacy of the Sichuan Opera to a wider audience through her compositions.

[1] Rashomon is a 1950 Jidaigeki psychological thriller/crime film directed by Fumio Hayasaka.

In *Dancing Upstream*, Pang includes the styles of Sichuan Opera, see Figure 15.15. These styles are conveying the emotional and dramatic content of this piece. The music plays a central role in the performance of Chinese Opera. All stage performances involve some form of singing and musical instruments. The principles of music in drama in Chinese opera will require a lot of physicality of the performers. It requires the efforts of a singer-actor who is a practitioner of Chinese traditional singing techniques and knowledgeable. Daily oral conversation is usually considered to be too realistic to Sichuan operatic style, thus the dancing and singing Sichuan Opera is one of the most influential forms of Chinese local operas. The tunes have various distinctive characteristics with rich local style. It mixes aspects of various Chinese musical traditions including Gao Qiang ( 高腔 , High-pitched Tunes), Hu Qin ( 胡琴 , Hu Instrument), Deng Diao ( 灯调 , Light Opera), Tan Xi ( 弹戏 , Tan Opera), Kun Qu ( 昆曲 , Kun Opera), all of which have contributed to the development of Sichuan Opera music. This genre influenced Pang's choice of harmony, melody, and percussiveness of the rhythmic accentuations of the piano.

Fig. 15.15: *Dancing Upstream*, Yan Pang and J-Sun, photo by Bill Cameron.

Pang examines the relationship and combination between traditional Chinese folk melodies and modern Western classical compositional techniques, including harmonizing and structuring music elements, which results in a hybrid cultural and musical exchange. She studies how people comply with the rules and break boundaries and investigate the practice of remaining flexible as a mode of inquiry. She uses these aforementioned strategies to inform her practice of improvisation in music and movement to add a unique touch.

Collaboration enriches traditional music because it centers on different time modalities. It all starts with the events that happened around us having

Fig. 15.16: Score part from *Dancing Upstream* by Yan Pang. The note heads don't not have precise values. The time management is up to the performers.

their own meaning. People can draw from these meanings in their own way, producing beautiful forms, full of the body of time. As previously mentioned Guerino Mazzola's collaborative work, tension and release create a dynamic and recursive composition through time, like a fractal structure.

Western classical music mainly relies on scores to enable the transmission of musical knowledge across time and groups of people. Composers are expected to make written works more efficient and performances more precise. Even folkloristic pieces throughout Europe have often been transcribed into scores to help the work of scholars and performers. However, there are many other musical contexts where the transmission of the repertory is mainly oral, and the production of music itself does not require written scores.

### 15.2.1 Left-Centric Interpretation

#### 15.2.1.1 Introduction

This chapter concerns left-centric string instrumental performance as an interpreter rather than a composer or improviser of new work. The emphasis is on the composition *Standing Alone*, for mountain dulcimer, by one of this volume's co-authors, Yan Pang [169]. Of particular importance is the transformation of the original version of the work—composed for acoustic guitar and entitled

*Luo* (Beauty in the Mist)—to the version for mountain dulcimer that bears its current title. Of most significance is that left-centric interpretation is so highly determinative of the perceived lengthening and slowed, expansive pacing of the work in its entirety, while its real time duration is significantly shorter than the original.

Although *Standing Alone* is written in traditional Western staff notation, left-centric performance practice dictates the actual durations of the most distinctive timbres. When occurring in mid-phrase, these are often permitted to vibrate for their natural, acoustically determined and acoustically evolving lives, as successive timbres are sounded. Cadence sounds—tones or chords, and typically distinctive timbres as well—are permitted their full durations and succeeded by ample, reflexive silences. The most important, time-shaping cadences are often comprised of a long-held timbre succeeded by a silence of similar length.

It is obvious both aurally and visually that Lubet's performance of this work is what he has characterized in the previous chapter as "left-centric," and to a greater degree than the guitar works discussed there. Lubet is entirely self-taught on this instrument and has largely developed both his technique and his repertoire—almost exclusively original works and works written for him—in isolation from the community around that instrument, whose primary interest is in Anglo-American vernacular tradition.

Like most mountain dulcimer players, Lubet uses only three strings, although some players/instruments use up to six. Having so few strings enables (but does not necessitate) a simple right-hand technique, using the index finger almost exclusively, though with important exceptions[2]. His left-hand dulcimer technique is even more intricate than that which he uses for guitar, as characterized in the previous chapter. On dulcimer, the left hand can take even more extraordinary responsibility than on guitar for every dimension of musical sound—pitch, timbre, and amplitude—which results in a shaping of musical time that is unusual in the repertoire of a quintessentially Western instrument. Perhaps surprisingly, this applies equally to those works written for him—and which, of course, he must interpret—as it does to his own compositions and improvisations. It is particularly true of Lubet's performance of *Standing Alone*.

#### 15.2.1.2 Interpretation as Creative Performance

Creative musical performancetypically encompasses improvisation and composition/arrangement, the latter either consisting of devising one's own part in an ensemble or making a work in its entirety. This chapter examines another form of creative performance, that of collaboration as a performer with composers to create new works. The creative component of the performer's role may be enhanced if the performer is also a composer, capable of providing valuable input from that perspective. An additional creative contribution from

[2] All but the fourth finger are employed at times.

performers may derive from their specialized knowledge of their instruments. The technique of certain instruments—such as mountain dulcimer—may be so unfamiliar and/or intricate to the composer that it may be necessary for the performer/collaborator to suggest substantive changes to the work in progress. This has indeed been the case with every work that has been composed for Lubet, to a varying but typically quite high degree.

Many composers have written new works for Lubet. At first, these works were for acoustic guitar, an instrument which presents formidable challenges to composers who do not play it, particularly when attempting to write polyphonically. In recent years, Lubet's principal instrument has been mountain dulcimer. Heretofore, almost no works in the "Western art music" tradition have utilized this instrument and most of the composers who have written for Lubet—nearly all of them Asian or Asian-American—were previously entirely unfamiliar with it, even as listeners, prior to having heard him play his own repertoire. Enjoying the sound of this rare truly American instrument, while also finding in it an affinity with their own Chinese, Japanese, Korean, and Indonesian musical cultures, these composers have made substantial contributions to Lubet's effort to build a "new music" repertoire for an instrument primarily associated with vernacular traditions. The characterization of this repertoire as falling within the Western art music tradition refers principally to its use of staff notation and the Western-style training of all the, mostly Asian, composers. The aesthetic, particularly of *Standing Alone*, leans East, as shall be discussed here. The dulcimer version of this work is more distinctive in regards to both interpretation—that is, the depth of influence of the performer—and its shaping of musical time.

#### 15.2.1.3 The Mountain Dulcimer

Among the challenges of writing for mountain dulcimer are its limited low range (D3-D6) and small number of strings. The typical fretboard (Lubet's included) is nearly diatonic—the Mixolydian mode plus the leading tone of the major scale, the latter added on as the instrument developed and arguably modernized.

While chromatic dulcimers are not uncommon, Lubet's choice of a non-chromatic instrument, still the norm, owes in large part to enjoying the challenges borne of its limitations, since he performs on several other fully chromatic string instruments. Further, although he rarely performs traditional dulcimer repertoire, intended for (nearly) diatonic fretboards, he is inspired by its folk character.

The non-chromatic version of the instrument presents a challenge but not an obstacle to performing chromatic repertoire. A broad palette of chromatic pitches (and also micro-tones) are available via a variety of tunings; pitch bending—with or without glissandi; natural and artificial harmonics; and playing on the left side (and both sides simultaneously) of the string. The time required to (prepare and) execute some bends, harmonics, and left side/both sides play—prominent in Lubet's playing and overwhelmingly non-standard techniques—mandate slower tempi, flexible pulses, and articulative silences. These devices are all left-hand techniques. In addition, the complexity and variety of these left-centric timbres and (micro-)tonal shadings demands that they be fully heard, that is, permitted to resonate their full, acoustically possible durations and thus display their diversity and timbral/tonal evolution over time. This foregrounding of acoustical richness is a fundamental aesthetic tenet of this performance, that is, of the composer's and interpreter's time sense.

#### 15.2.1.4 Left-Centric Play and the Mountain Dulcimer

Lubet is completely self-taught on mountain dulcimer, having prepared himself via decades of performance on various guitars, basses, and ukuleles. There are, however, major differences in nearly every aspect of technique between these and dulcimer.

Indeed, Lubet's only original intention in learning dulcimer was to train himself mentally to play a string instrument that was in important ways opposite from the others he knew in both hand position and the order of its strings. (Notably, there is a somewhat conceptually related Confucian tradition in various Asian cultures of plucked string instrument repertoires originally intended for meditation rather than public performance.) Consequently, Lubet has developed a strongly left-centric dulcimer technique different from any other player.

In addition to using the same left-hand techniques described in the previous chapter as applied to guitar, on dulcimer Lubet uses his right-hand index finger almost exclusively, in opposition to—and less elaborate than—the (multiple) fingerpicking or plectrum use favored by other modern performers. Although largely reliant on a single right-hand finger, he uses both up- and downstrokes. While alternating strokes are the norm for plectrum playing, Lubet is unaware of any other performer using them as a finger technique on any string instrument. Using a finger this way alternates the different timbres obtained with the fingertip (upstroke) and fingernail (downstroke). The finger downstroke is used in Western music only for certain strumming techniques, including those associated with flamenco guitar. Used on individual strings, the finger downstroke is a technique Lubet acquired from the Chinese lute, *p'ipa*. It is especially apt for the natural harmonics so appropriate to ma [157, p. 26], [114].

More broadly, though, Lubet's repertoire evokes another Chinese string instrument, the *qin*. The affinities were first noted by Yan Pang, but also by her Korean colleague and gayageum virtuosa Soojin Lee, both of whom referred to Lubet as a "qin master." The qualities of *qin* music are exemplified in the performance of master Chen Leiji [21]. Resemblances in Lubet's guitar compositions and performance can be heard in the album *Spectral Blues: New Music for Acoustic Guitar* [40]; several pieces from this recording referenced in the previous chapter. The similarities are neither studied nor coincidental.

Lubet has been intrigued with *qin* since working in Taiwan in 1993, but did not deliberately emulate it.

It is evident both audibly and visually in Maestro Chen's performance that *qin* requires highly nuanced left-hand technique for tuning (*qin* has no frets), fast playing that uses a wide variety of what Western string players call hammer-ons and pull-offs, embellishments (including glissandi), and natural harmonics. These are all elements of Lubet's left-hand technique on his instruments as well. At the same time, the demands *qin* technique makes on the right hand [156] are even more formidable. The Ming Dynasty treatise *Taiyin Daquanji* (Great Collection of Superlative Sound) describes 75 right-hand techniques and 'only' 49 for the left.

As a performer who is mostly self-taught on some instruments and entirely self-taught on others, Lubet has not characterized and catalogued his techniques in the kind of granular detail seen in *qin* treatises, though they are surely many and nuanced. The relative simplicity of Lubet's right-hand dulcimer technique on just three strings, results in most of the 'action'—notably including elaborate ornamentation and a large repertoire of timbres, mostly natural harmonics and multiphonics—residing in the left hand, in playing that is left-centric, rather than left-handed. As noted in the previous chapter, left-centrism lends itself to legato events, with emphatic time markers less frequent than in the kind of right-centric playing that features fast plucking/picking of individual tones/chords. More important, there is a vastly expanded repertoire of timbres, especially high, mostly natural harmonics, up to and including the 11th partial, whose raisons d'être are to be appreciated as sound objects in and of themselves, a purpose that demands that they be allotted the full duration dictated by their unique acoustical properties and often followed by silences, sometimes of substantial length, both to allow for reflection and to void the possibility of a connective relationship to a subsequent sound having priority over the sound itself. When timbre is thus foregrounded, duration becomes a function of acoustical properties, the natural lives of diverse timbres, rather than the articulation of pulse. The use of extended silence to suspend measured time also serves to disconnect individual acoustical events and in doing so renders them the fundamental expressive units, rather than relationships, the smallest of these being intervals, whether comprised of focused, stable pitches, or, in instances of more complex sounds, of a more anharmonic character.

#### 15.2.1.5 East-West Aesthetics of Musical Time: Exchanges and Collaborations

The late Chinese-American composer and ethnomusicologist Chou Wen-Chung (1923-2019) believed that the most important distinction between Western and Asian musics was that the former's smallest unit of expression is the interval [24, pp. 214-18], the latter's the single tone. Although Lubet's time with Chou consisted of only a couple of guest lectures and a single private lesson in 1975, during his graduate studies at the University of Iowa, he regards him as his most influential composition teacher, particularly as concerned his ideas about single sounds. Given that the expressive force of single tones requires that these must at times be long enough in duration (and sufficiently exposed in texture) to fully express themselves, sometimes very long indeed. Further, these tones should at times be preceded and/or followed by silence, in order to permit reflection upon their aesthetic qualities. They often benefit from sparse textures in which the evolution of a tone over its (often long) life can be clearly heard without the (Western) distraction of dense polyphony. Western classical music notation only rarely indicates and struggles to describe the highly nuanced evolution of single tones over their lifespans, in subtle variants of pitch, timbre (sometimes including orchestration), and dynamics. While good—creative—performers may indeed attend to these, their rarity in notation indicates that they are not typically a priority in composition.

In the documentary Itzhak [22], violinist Itzhak Perlman conducts a class (28:38-31:59) in which he and his students are listening to and comparing four performances of Joseph Joachim's arrangement for violin and piano of Brahms' Hungarian Dance #17 in F-Sharp Minor. At 31:00, one of his students proposes that the 'front' and the 'end' of a note are as important as its 'core' to its emotional content. Perlman concurs with and expounds upon this idea, but what is of greatest interest is that this kind of granular analysis of the micro-levels of time where key facets of musical expression reside within performances of the Western canon reveals that this 'cellular' manifestation of the body of musical time is a major preoccupation of (the best) performers and only rarely a compositional consideration. The exchange of ideas in this master class challenges conventional wisdom about the division of creative labor in the realization of works of Western classical music, giving truly thoughtful performers the credit they are due. Perlman goes on to acknowledge the 'micro' aspects of string performance that he treasures as 'esoteric' and 'inexplicable,' though these appear to be analogous to the techniques the abovementioned *qin* treatise explains in very precise gestural rather than temporal detail.

In plucked string instrument performance, the dominant (typically right) hand's responsibility is, with few exceptions, limited to the initiation/attack and, to prevent a sound from sustaining for its full natural value, dampening. Any variants during the life of the tone are thus the responsibility of the non-dominant (typically left) hand. A left-centric plucked string music is thus one whose aesthetic emphasizes the expressive character/aesthetic nature of long single tones. The oft-held but generally incorrect notion that the role of the left hand is more important/more challenging in most string music may thus actually be correct in left-centric repertoire.

#### 15.2.1.6 Teachers, Students, Collaborators

The aforementioned Chou was a celebrated composer and scholar, among the most important American composition professors, and the leading disciple of French-American composer Edgard Varèse. Chou's interest in Varèse arose from the Varèse's aesthetic of individual sounds that, conceptually, though not stylistically related to Chinese/Asian musics, Chou saw as an important affinity [24, pp. 214-217]. The composers had an enduring relationship, beginning as student and teacher, with Chou eventually becoming the executor of Varèse's artistic estate, a responsibility that including completing several works unfinished at the time of Varèse's death.

Lubet's relationship with Yan Pang, also began as teacher and student and has also been multi-dimensional across a wide variety of artistic, scholarly, and professional collaborations. The parallel of this relationship with that of Chou and Varèse is obvious and, like that partnership, begin with a common musical affinity for sounds themselves rather than connections between sounds. The difference, however, is that the Lubet-Pang affinity is as stylistic and East-leaning as it is conceptual. While Lubet's formal study of Asian musics was limited to a brief course in South Korea in 2014, he has been listening, traveling, and collaborating across cultures since high school in the 1960's. Pang's interest in Lubet's work, especially the prominence of complex sounds and silences, both of long and unmeasured duration, which she heard as evocative of the concept of ma in Japanese music [157, p. 26] and related aesthetics in other Asian musics as well.

Lubet has not catalogued his techniques in detail, though they are many and nuanced. (Notably, Itzhak Perlman's, in the abovementioned master class, implicitly acknowledges that Western classical violinists have not done this, either. Indeed, he questions whether it can be done at all.) Pang compares Lubet's performance style to the repertoire of the ancient Chinese *qin*, which is, by contrast, a highly schooled tradition with treatises that catalog and describe the instrument's many techniques for both the left and right hands.

#### 15.2.1.7 *Standing Alone*

*Standing Alone*, originally entitled *Luo (or Beauty in the Mist)* was originally composed for acoustic guitar by Yan Pang in 2013, early in her studies with Alex Lubet. As is the case with many/most works composed for a guitarist by a non-guitarist, the depth of collaboration was rich and replete with exchanges of ideas. These often concerned the most typical challenges to the non-guitarist composer, among them the voicing of chords and the availability of harmonics. Designated "Largo ad lib" without a metronome mark, the performance is almost glacially slow and, as per the composer's instruction's, often rubato and abounding in space, at times owing to the technical demands of getting from one complex sound to another.

In 2017, Lubet created a new version of *Standing Alone* (with its new title) for an all-Chinese solo dulcimer program performed on television in Chengdu, Sichuan Province. The program consisted of works by Chengdu composers, all but one originally written for other instruments (acoustic guitar, classical guitar, violin, and *p'ipa*) and Lubet's own *At a Place Far Away*, a fantasy on a well-known folksong from Southwest China.

In an exceptional gesture of confidence in collaboration, Pang entrusted Lubet with creating the new version of *Standing Alone* without further exchanges or interventions. The most obviously transformative challenge was the reduction of strings by half, from six to three. Octave doublings were eliminated as were dense harmonies. The latter especially had the impact of pulling the style away from Western classical tradition and closer to both *qin* and the folk tradition of the mountain dulcimer itself, with its favoring of melodies supported by drones. These meeting of these style is, perhaps surprisingly, compatible.

The texture of the dulcimer version of *Standing Alone* is thinned from the guitar version not only by halving the number of available strings and losing almost an octave of low register, but also by Lubet's radically left-centric dulcimer technique. As Chou has argued, albeit in different words, a minimally polyphonic texture such as is heard here (and in some other repertoires, such as that of Japanese *shakuhachi* flute, a monophonic texture) affords single tones/chords/sounds the ability to function as lowest-level expressive units, their complexity unobscured by relationships with other voices.

In transforming from guitar in *Luo* to dulcimer in *Standing Alone*, nothing is added. But "nothing is added" must be construed in two ways. There is, as composers and especially composition students are fond of saying, no new material. (This author's dislike for referring to music/sound as 'material' is exceeded only by his distaste for characterizing musical relationships as 'horizontal' and 'vertical.') But there are newly added 'nothings'—that is, silences—that emerge organically from a more apparent aesthetic of individual sounds, as well as from left-centric play. The newly added 'nothings' in *Standing Alone* only rarely need to be notated by the traditional use of rests. Because successive notes/chords/sounds/expressive units are so often timbrally/dynamically distinct, legato, as it is conventionally understood, cannot occur. In addition, articulative and sometimes also reflexive silences are an inevitable acoustical/physical/technical/aesthetic consequence of advancing from one expressive unit to the next. Even with a paucity of conventional legato, though, flow pervades, because virtually every sound is sustained as long as possible, either permitted to vibrate for its natural acoustic life, especially in the case of harmonics, or until a note/chord/sound is replaced by another on the same strings.

At the same time, an important psychoacoustical phenomenon occurs when the instrumental texture is thinned into expressive units that are perceived as more multi-dimensional, more meaningful than simply "notes" or "chords." The effect of changes of pitch, even within very narrow ranges, whose sounding different from each other is something 'more' than simply as different notes, but also substantial if subtle transformations of timbre, often enriched by changes of amplitude. An exemplar of this phenomenon is Density 21.5 for solo flute, by Chou's teacher, Edgard Varèse. A 1998 recording by flutist Jacques Zoon, under Chou's supervision, can be heard at [172].

The move from guitar in *Luo*, with a relatively normatively active right hand, to radically left-centric dulcimer in *Standing Alone*, and thus from harmony to a texture close to monophony, while thinning acoustic density, thickens acoustic meaning, as every aspect of each sound—frequency, register, timbre, amplitude—attains larger, expressive value as a facet of the sound being heard as an entity rather than a link. These expressive units are characteristically enforced in their autonomy by interceding buffers of 'nothing' that ensure an aesthetic whose foundation is their unitary rather than relational value.

The left-centric palette of timbres and tonal inflections used in both versions of *Standing Alone*, but especially the dulcimer version, are largely those heard in Lubet's own guitar work, as described in the previous chapter. But what is more interesting is the evolution of time shaping and time perception in the move from larger to smaller instrument.

When Lubet re-created *Standing Alone* for mountain dulcimer from *Luo* for acoustic guitar, it was an individual act, done at composer Pang's request but without her intervention. She approved the finished work as delivered—based on both the notated score and Lubet's performance—without requests for revision.

One of Pang's observations about the new work, whose transformed character was of such magnitude as to regard it as a truly different creation from the original—concerned its duration. Pang commented that Lubet's performance of *Standing Alone* was twice as long as she'd imagined. While Pang described this phenomenon in real time durations, it is her sense of temporal proportions that is of greater significance.

Of utmost importance is that Pang wholeheartedly approved of a performance of her work that she perceived as twice as long as she'd imagined. In real time, though, the performance was actually significantly shorter than her estimate. Her original conception of the dulcimer version's duration was based on her recollection of the—presumably far shorter—guitar version. However, in reality, the guitar performance is the longest of all versions, real or imagined, and as long as Pang believed the dulcimer version to be. The actual dulcimer version—that is, Lubet's performance, as sanctioned by the composer and perceived to be longest by far, was significantly shorter.

The above is in no way intended as a psychological test of Pang's ability to estimate real time durations of either notated scores or actual performances, nor is it meant to be an n=1 pilot study of real time perception. For this reason, real times are not provided here. Rather, what is significant is that the dulcimer version of *Standing Alone* is experienced as temporally expansive for reasons deriving from the character of its transformation from *Luo* the guitar original.

The guitar version is both more harmonic and technically relatively conventional in its distribution of 'action' between the hands, in large part owing to the use of thick chords. The sense of harmony creates a more Western, relational aesthetic, in which sound bodies are mostly valued for their succession or linkage, rather than their being, functioning as the circumference of intervals, as Chou characterized them in his Iowa lectures. In such a sonic environment, successive frequency ratios prevail in terms of meaning and character over all other sound dimensions, such as timbre, tonal shading, and amplitude, these functioning as a kind of acoustic paint. The degree to which the *Luo* is Westernized by harmony should not be exaggerated but neither can it be denied.

In *Standing Alone*, nearly monophonic and never truly harmonic or contrapuntal, the depth of detail in individual sounds, especially at formal junctures, makes or perhaps enhances a more Eastern aesthetic of distinct expressive units. Silence is required to separate sonic objects to confirm their value are individual entities. But in many instances, the sounding of the expressive unit for its full possible duration and the totality of its acoustical richness also demands that it be allowed to die a natural death, its completion in sonic decay only consummated in silence. Thus, silence—that is, nothing—is sometimes an articulation. At other times, it is the inevitable culmination of a complex, if increasingly barely audible, expressive unit.

The perception that *Standing Alone* for dulcimer occupies substantially more time than its parent work, *Luo* for acoustic guitar, derives not from its real time duration but from an aesthetic that is unambiguously grounded in the value of individual sounds, rather than ratios between successive sounds, as the fundamental expressive unit. Thus, *Standing Alone* is the longer work by far in 'event time,' with its greater abundance of both individual unit events and varieties of silences: silences that demarcate events and those that culminate events. *Luo* approaches this aesthetic, but is called away from it by moments of harmony that locate the work somewhere between East and West, between sounds and ratios. The migration to dulcimer in *Standing Alone*, played with a radically left-centric, largely monophonic technique, rarifies the texture, while thickening the meaning of notes, transforming them into acoustically rich events, temporal locations, rather than linkages to what happens next. As a work that treasures the autonomy of bodies of time, it is most fitting that *Luo* became *Standing Alone*.

### 15.2.2 Performer Time/Composer Time

The previous chapter refers briefly to a class taught by violinist Itzhak Perlman as part of the 2018 documentary American Masters: Itzhak [22]. Although the excerpt of the class is brief (28:38-31:59), it has much to say about the manner in which, in Western art music, both the perception and the shaping of time differ radically between performers and composers. The difference is especially significant for violinists and other bowed string players, who exercise extraordinary control over the lives of the sounds they make throughout their durations. Examining the difference between "performer time" and "composer time" mandates a consideration of what is meant by culture, distinctions between cultures, and ultimately that culture is more time than place.

In the class Perlman is teaching, he and his students are comparing four recorded performances of Brahms' Hungarian Dance #17 in F-Sharp Minor, in arrangements by violinist Joseph Joachim and Fritz Kreisler. The performance heard in the documentary is by Yehudi Menuhin [110] arranged by Joachim [62]. Perlman praises two others, Fritz Kreisler [74] and Jascha Heifetz [56], both in the arrangement by Kreisler. (The fourth is never identified.) For the purposes of a chapter that democratizes a creative process that typically over-valorizes the role of "the composer," the musical selection in question could hardly have been more apt. The large majority of the Hungarian Dances, including #17, are, rather than original compositions, settings of "gypsy" melodies, originally scored for two pianos. The works actually being discussed, though, are violin and piano arrangements, by different violinist/composers (with Kreisler playing his own version), rather than by Brahms himself, the latter having appropriated the melodies mostly from folk sources. Since, in addition, Perlman and his students have important things to say about the highly nuanced differences between the violinists' extravagantly personal but also distinctly "Eastern" interpretations (the Viennese Kreisler's somewhat less so), it is evident that, even in a repertoire and tradition that often worships "the composer's intentions," creativity is a larger and more evenly distributed—if not also fairly acknowledged—process.

In less than four minutes of documentary film, Perlman and his students engage in a profound discussion of performance that, perhaps inevitably, focuses upon time. The Heifetz rendition is the focus. As Perlman observed:

> And the Heifetz one, I felt awestruck by the middle section, you know, what he did with colors. And it was almost showing off where it was just like one color after another and it was just came (?) with you it was like it was like fireworks boom, boom, boom, boom, boom, and it was all perfectly timed and so on...

In response, a student asks:

> Do you think timing could affect the illusion of how we perceive the sound?

The music in question is replete with repeated chords, shaded with subtly different timbres. The contrast is most noticeable at phrase endings, but, as the young student observes, made more apparent by timing, with a pulse that is in constant if mostly subtle flux. Perlman's bright and inquisitive student almost certainly knows the answer to his own question (which may have fallen victim to editing), itself a reimagining and amplification of his teacher's observation; the perception of contrasting timbres is in a necessary symbiotic relationship with an organically fluctuating rather than mechanical—and strictly notated—pulse.

Another student remarks (possibly after remarks edited from the film?):

> The front and the end of the note affect what you're hearing and how you emote to it. I think front like, someone was talking about the bowing and how and how hard it was. I think articulation and the way you end notes is just as important as the note and the core of the sound

itself. Cause if the core is gorgeous but it ends totally flat, emotionally you're gonna, it's, it's not gonna (unintelligible)...

The entire exchange about Heifetz' performance concerns "color"—that is, timbre—and timing and their interrelationship, in the context of single tones/chords. There is both a regard and an ear for a level of nuance, complexity, and sophistication far beyond what the performers received from composers and their traditionally notated scores, borne in large degree from an intimate knowledge of their instruments, far beyond that of non-composer performers. (Even Kreisler, performing his own arrangement, did not—and likely did not wish to—capture the intricacies of timbre and timing of his playing in his score.)

This collaborative analysis between teacher and students went even further than individual tones. The second student shared her idea that emotiveness is dependent upon granular control of tone production; that notes are comprised of "articulation"—by which she means attack—"core"—or sustain—and "end", that is, release.

This remarkable conversation revealed that elite Western classical violinists hear their performances as comprised of individual, granularly conceived and executed tones as the fundamental expressive units of their performative art. This is not only a very different conceptualization and scaling of time than the notated repertoire that has been supplied to them by composers, it bears an extraordinary resemblance to Chou Wen-Chung's characterization of the foundational role of individual sounds in Asian music in general and to the intricacies of tone production as described in *qin* music treatises in particular, as discussed in the previous chapter. From this, it is possible to infer that, in Western classical music, performer time and composer time differ radically. Further, it may be that Western and Eastern performer time have much in common.

#### 15.2.2.1 Coda: Culture and/is Time

The observation that performer time may have a common basis across at least some cultural traditions mandates even closer examination of every element of Perlman's class.

The repertoire in question, the Brahms Hungarian Dance #17 in F# Minor is intriguing, not only for being an arrangement of the original composition for two pianos—actually two different arrangements, by Kreisler and Joachim—of a setting, in four different interpretations, but because the source material for Brahms' composition is a melody of Eastern European origin. Although by one standard Brahms is the most archetypally Western of classical composers—he is the last whom archconservative music theorist Heinrich Schenker believed to belong to the great tradition—the Hungarian Dances are evidence that Brahms "learned the language from those who spoke it as a mother-tongue" [13, p. 237]. The performance by Heifetz, a Russian-American Jewish immigrant, is so rhapsodic and so much more rhythmically intricate than the score, that, in the right venue and perhaps a different kind of attention that a "classic" supposedly deserves, it could easily be taken for a folk performance, rather than a "masterwork"" by Schenker's poster child.

Given that Israeli-American Perlman has chosen recordings of Heifetz, Menuhin, and Kreisler (the latter of Jewish heritage, but baptized) it might even be argued that, in contrast to Western classical composition, classical violin performance is a largely Jewish or Eastern tradition. (The Jewish contribution to Western classical music performance, of course, extends far beyond Perlman and his CD collection.) The Heifetz performance in particular reveals performer time far more intricate and arguably considerably more Eastern than the composer time of Brahms' score, whose notation system is far from culturally neutral in regards to nearly every musical dimension.

That Perlman's studio and "Western" classical music in toto are largely populated by musicians of Asian heritage is another data point in favor of cultural fluidity. The impact of Eastern influence on Western composition is centuries old and has been much enriched by participation of many Asian and diasporic Asian composers. But it is almost certainly the case that Western classical performance practice has not been influenced by (diasporic) Asian musicians. As both Yang [168] and Wang [163] have observed, the impact of the introduction of Western classical music into Asian life in general and Asian music education in particular, where it has been strongly conflated with modernization, has been to diminish the presence and status of indigenous musical traditions.

Thus, the similarity of performer time across cultures, perhaps especially in string performance, seems to owe to something else. Perhaps it is a common ethic/aesthetic of responsibility for the sounds throughout those emotive units' lives, in a manner typically lacking in Western composers, who in general have a removed relationship from performance, even when they are performers themselves, as mandated by a system of notation that is culturally biased toward pitch relations and rhythmic symmetry.

## 15.3 Christopher Rochester: Manipulating the Subconscious Gravity in Musical Time

**Summary.** We have focused a great deal of our attention on understanding different perspectives on time and their relations with musical time. With our newfound understandings, we have built a stable foundation for which to further explore new time related relations. Christopher Rochester has conceived an investigation that aims to manipulate a musician's execution of time through subtle alterations of a controlled variable. If successful, we may be able to begin developing formulae that are able to partially quantify the subconscious variances in group playing.

– $\Sigma$ –

What is the underlying reason (excluding race...[3]) that we do not hold rhythmic theory to the same standards of importance as harmonic and melodic theories? This answer can be found within the genre of classical music itself, a style of music that relies on the ictus of the conductor and is composed primarily of string instruments. This is somewhat of a generalization, but classical musicians often react to the pulse of a piece, and jazz[4] musicians often internalize the pulse of a piece. Orchestral musicians react to the visual representation of the pulse given by the ictus of a conductor. As they do that, they then divide rhythms accordingly to said visualization. Jazz musicians, however, rely on a concept known as *phase locking.* This concept finds its foundation in the idea that different musicians will collectively attempt to play in sync in order to create music. This problem is not as apparent in orchestra, because of the nature of string instruments. The initial attack transient of string instruments is gradual in comparison to many of the instruments used in popular styles of music. With the initial attack transient being gradual, there is less of a demand for them to play in complete sync with one another. Stringinstruments have the luxury of being a fraction of a second out of sync, but still give the illusion of being completely together. This same luxury is not afforded to most styles of music involving drums and other instruments with percussive qualities. With more distinct initial attack transients, instruments common in popular American music must phase lock in order to sound *good.*

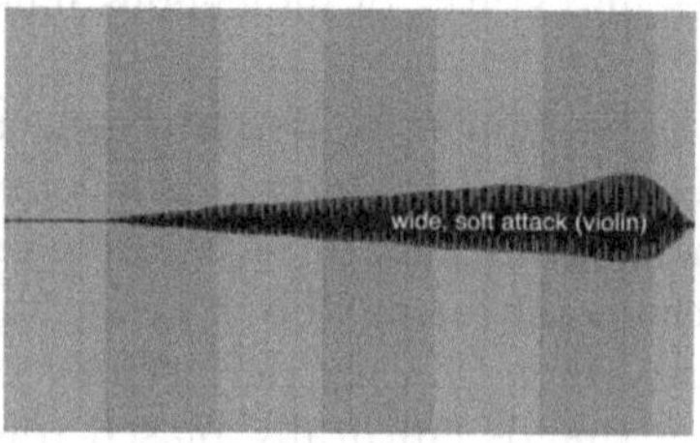


Fig. 15.17: Wide, soft attack (violin).

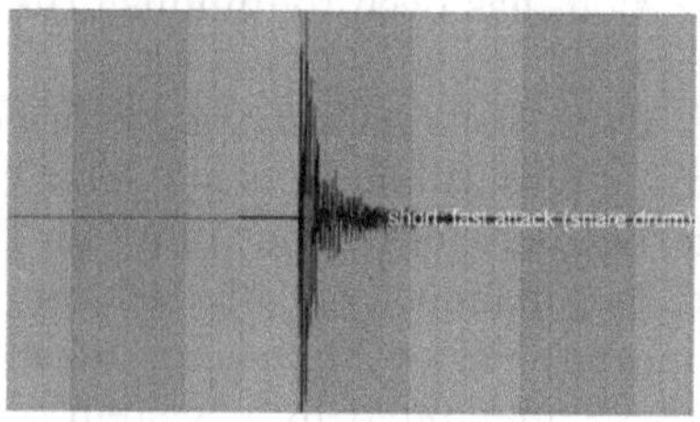


Fig. 15.18: Short, fast attack (snare).

Without an imperative need to discuss phase locking in the orchestral setting to perform at a high level, particular concepts of rhythm have gone untouched in music education. We should further our understanding of rhythm regardless of its perceived absence in certain styles of music.

[3] The importance of the education of music by people of color has long been an issue in the Western canon, as they continue to focus on the achievements of a eurocentric way of thought.

[4] Jazz in this case refers to the most popular styles of music in the US.

### 15.3.1 Gravity in all Music Theories

There is a felt gravity in all theories of music, though it might not necessarily be referred to directly. The idea of gravity in music stems from its own existence in reality,

"Gravity is the force that attracts a body toward the center of the earth, or toward any other physical body having mass." – Merriam-Webster.

If we use this definition and apply it to music, we can see there are the same occurrences in all divisions of music theory. Imagine tossing a ball into the air, and shortly catching it thereafter. This is similar to a visual representation of a chord progression. The ball begins in your hand, which we can equate to the I chord in music. The act of the ball moving upwards might equate to the idea of a subdominant chord, or the IV chord. It then begins to fall, with the same attitude of the V chord wanting to return to its starting location (the I chord). Further, the ball can be caught by another hand creating the visualization of the concept of a modulation. The important idea to absorb in this visualization is the movement of the ball. Certain chords are linked with certain motions. V7 chords embody the idea of gravity, as they want to resolve.

The same concept of gravity is apparent within melodic theory as well. Equal temperament and the division of notes in an octave create fundamental functions of melodic notes in context. Do, Re, Mi, Fa, So, La, Ti.... We have quickly created an instance in which Ti, the leading tone, has been subjected to the influence of the gravity of Do. Ti *wants* to go to Do, whether or not it goes there is an entirely different subject. The idea of gravity further gives freedom from certain constraints we have been introduced to in Western music education. In Japanese styles of music, pitches are altered to create a sense of gravity, because the gravitational center is paramount in the music's identity. b2 can be even flatter due to the nature of it *wanting* to resolve to 1, and the leading tone can be sharper in many instances due to the nature of it *wanting* to resolve to 1 as well.

The idea of gravity in both harmony and melody was introduced to reinforce its importance to music. This idea of gravity should be considered heavily influential in our understanding of rhythms in music. Consider that most scales are not divided equally within an octave, but we often attempt to divide our ideas of rhythm equally. This equal division of rhythm will often lessen our understanding of the rhythms of music we are not familiar with, and further create a tendency for musicians to become deaf to the certain rhythmical nuances in music. If we can further the discussion of gravity in rhythm, we can understand that the importance of gravity outweighs the importance of the depiction in notation.

### 15.3.2 The Black American 8th Note

Jazz, blues, swing, hip-hop, soul, and many styles of music born from Black American communities share hidden complexities within the very fabric of their being. Our current understanding of time in music has created an opportunity to recognize and interpret the feel of many styles of music. Particularly, the minute differences in musical timings that create the concepts of feel and groove. Popular music in 4:4 time tends to have many genre overlapping characteristics. These similarities allow us to focus on particular differences in rhythm. As many elements of harmony and melody remain constant throughout these differences in genres, we now have an organization of rhythm in which we can focus on certain information we find important.

As we now have an understanding of our choice of 4:4 time in certain Black American Music styles, let us begin to explore a particular example. We will first begin with learning about concepts of swing in music.

The term *swing* comes from the idea of an uneven organization of 8th notes in subdivision. As we have discussed previously there is a certain gravity in rhythm that helps form the basis of a socially acceptable *feel*[5] in the style performed. Though these styles obviously vary, let us speak to the more widely accepted understandings of the organization of *swing*.

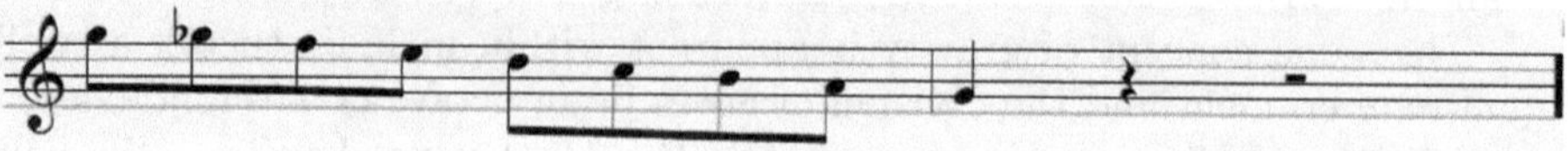

Fig. 15.19: Example 1.

In Example 15.19 we see is a simple depiction of the descending G7 Bebop scale[6]. Note the melodic implications adding that the note (Gb) created melodically and rhythmically. The chord tones align with the strong beats of 1, 2, 3, and 4. The emphasis on down beats allow for a balance of even and uneven organization of notes.

Let us further focus on what we are actually viewing in example 1. 8th notes, as they exist, tell us that the quarter note is divided symmetrically in half. Each 8th note is valued at .5 value of a beat, whereas a quarter is valued at 1 value a beat. The truth, however, is that we more often than not do not divide our beats evenly as notated. These variations in the distribution of weight created characteristics of feel that are apparent in all styles of music. The amount of gravity in a piece will often lead to certain qualities in performance that are often overlooked. Realize that though example 2 and 3 are notated differently, musicians will often think of the notation of example 1 and play the rhythms of the other examples.

Example 15.20 refers to a style that is akin to *swingin'* in the style of Ragtime, marches, and many styles of music. The melodic and harmonic information has remained the same, but there is a substantial difference in the

[5] Feel refers to how others perceive the success of the organization of your rhythm.
[6] 1 2 3 4 5 6 b7 7

Fig. 15.20: Example 2.

rhythmic communication of the idea. If we break it down further, we realize that the 8th notes are now grouped into a further subdivision of 16th notes (4 notes to a beat) with an uneven distribution of 3-1. This 3-1 distribution results in a different feeling due to the influences of gravity in rhythm. Just like the Japanese Shakuhachi player would alter the pitch of a melodic line to emphasize certain points, the rhythmic organization in swing has achieved the same results. The amount of distribution varies in style, but we have come to our first representation of a group of notes that both function and do not function as 8th notes.

The subdivision of 4 is fairly common, but it lacks a quality that is found in other subdivisions. That quality stems from the ability to divide 4 evenly, but what happens when we divide subdivisions into prime numbers. This concept of subdivision based on primary numbers has largely ingrained itself into Black American Music. Realize, that this choice of subdivision is both a conscious and subconscious effort. That is to say, creators of music did not always aim for a particular grouping of subdivisions, rather they intended to aim for something that felt good to them as listeners.

Fig. 15.21: Example 3.

Example 15.21 is largely associated with medium swing in the jazz idiom. The focus has shifted to an uneven distribution of 3-1 (ragtime, marches, etc...) to a distribution of 2-1. This style has become a focal point in the teaching pedagogy of jazz educators to explain where an improviser should start thinking rhythmically. There is a rhythmic tension that is created through this prime distribution of metric division. 8th notes are the most common rhythmic device of improvisers, which has a value of .5 of a beat, further meaning two 8th notes will fill the space of a quarter note. 8th note triplets have a value of 1/3 of a beat, further meaning three 8th notes will be the space of a quarter note. This creates an understood interpretation of replacing the understood 1/2 division of a beat into the uneven 2/5—1/3 division of a beat. Which might sound wrong, but if we think back to the idea of repetition, this creates the foundation of the development of feel in popular music to come in the US.

Be it gravity, especially the uneven distribution of 8th notes, there is a need to openly discuss the implications of the concepts in music education. Allowing us to have a greater grasp on our own interpretations of music will give musicians a better opportunity to accurately participate in styles of music outside of their own. Though these two concepts are merely fundamental ideas, the implications of absorbing them will allow people to discern differences in music to which they have previously been blind. This style of analysis will further create a means to communicate what they hear in an academic setting, which can be further used as a tool to create new music, not as tightly bound by the lack of knowledge of theories of rhythm.

The idea that something can and cannot function as something deals with certain flaws of our notation system. Though we can accurately depict rhythm in western notation, there comes a point in which it becomes exceedingly complicated. If I were to ask you to cut a cookie into three equal parts, any attempt would leave some form of an inaccuracy in execution. Intention often varies from execution, which means though we might refer to a distribution of 8th notes as a 2 - 1, we might be really listening to a division of 30 - 17 with a 47 note division of the quarter note. This concept aligns with Charles Keil's, *The Theory of Participatory Discrepancies: a Progress Report.* Participatory discrepancies introduce the concept that no two musicians divide their 8th notes exactly the same, which is not a farfetched statement. But, realize that though the execution does not always match the intent, the intent should be acknowledged when it comes to execution. There is not an amazing amount of research that deals with participatory discrepancies, but it begs to create new questions. Do musicians prefer to play with particular musicians because their subdivisions of rhythms closely align with others? The short answer would be yes, and this is why musicians must alter their conceptual divisions of rhythmic understanding to suit the appropriate style of music. 8th notes in rock function very differently from 8th notes in funk music, and musicians will more likely than not, prefer musicians that participate in phase locking.

As we begin to explore the divisions of rhythms in 4:4 Black American music, let us now switch genres again. We will now look into an example of Neo-Soul, a style of music heavily influenced by the contributions of J Dilla[7]. One of J Dilla's most influential contributions stems from his use of samples in music. The samples he used often did not perfectly align with the music he was creating, these uneven sample lengths only reinforce the concepts we have previously discussed. Through repetition, our understanding of the distribution of 8th notes has evolved.

Erykah Badu's *Fall In Love* is a great example of the different understanding of subdivision in 4:4 music, see Example 15.22. Understand though, that even though this can be illustrated with a division of quintuplets, the rhythm

[7] An American record producer and rapper who emerged in the mid-1990s underground hip hop scene in Detroit, Michigan, as one third of the acclaimed music group Slum Village.

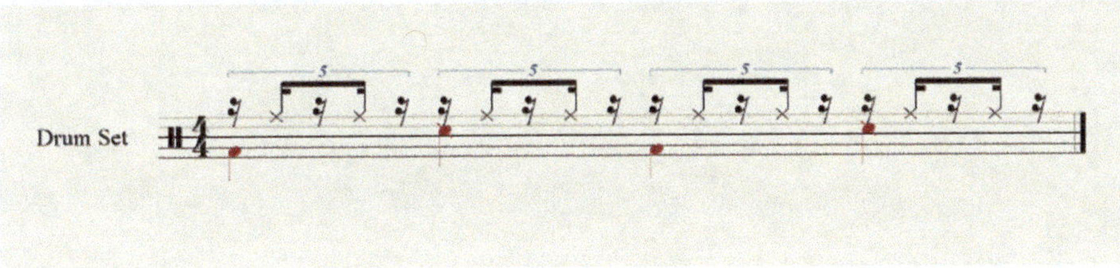


Fig. 15.22: Example 4.

section is not necessarily thinking of this subdivision. It is more like that musicians have mutated their understanding of subdivision to coincide with this quintuplet division of the quarter note. This distinction allows performers to both understand this music can be realized as a subdivision of 5 and/or the standard subdivision of 4.

### 15.3.3 Conclusion

Be it gravity, especially the uneven distribution of 8th notes, there is a need to openly discuss the implications of the concepts in music education. Allowing us to have a greater grasp on our own interpretations of music will give musicians a better opportunity to accurately participate in styles of music outside of their own. Though these two concepts are merely fundamental ideas, the implications of absorbing them will allow people to discern differences in music to which they have previously been blind. This style of analysis will further create a means to communicate what they hear in an academic setting, which can be further used as a tool to create new music, not as tightly bound by the lack of knowledge of theories of rhythm.

# 16

# Experiments with Local and Global Rhythms

**Summary.**
Here the theory of local and global meters/rhythms is applied to empirical investigations about the statistically significant correlation between analytical data (harmonic, melodic, rhythmical analyses) and the measured micro-timing (agogics).

– $\Sigma$ –

This section deals with difference between the symbolic time in scores and the physical time quality generated in concrete performances. This difference is not just a formal topic, since the shaping of time generates a musical ontology that expresses the semantics of time in the musical reality of performance. Musical time here is described in terms of tempo curves. Such curves express the shaping of time as a trace of analytical insights, i.e., the loading of time by contents of a detailed and musically significant nature.

The detailed description of our experiments is given in [95, Part XI]. In this section we stress the semantic aspect of these analyses, rather than repeating those procedures. The examples we refer to deal with the following compositions: Schumann's op. 15/7, *Träumerei*, and Chopin's *Prélude* op. 28/No.4. For both compositions, we analyzed 28 famous performances for Schumann's composition and 30 famous performances for Chopin's composition. The raw data were given in terms of tempo curve measurements.

The analysis of these data was shaped under the hypothesis that such tempo curves are expressions of harmonic, melodic, and rhythmic analyses of these compositions. The result states that tempo curves are given by an operator, called *Beran operator*, which states that the logarithm $ln(T)$ of tempo $T$ is given by the formula

$$ln(T) = \Omega_\omega^X + C$$

where $\Omega_\omega^X = (X, \omega)$ is the shaping vector with $X$ being an onset $E$ function $X(E)$ of the specified analysis, and $\omega$ is the *shaping* vector of random variables. The theory states that a good choice of $\omega$ yields the above formula. This state-


G. Mazzola et al., *Making Musical Time*, Computational Music Science,
https://doi.org/10.1007/978-3-030-85629-8_16

ment is a statistical statement, and it is statistically significant, the details are, as said, explicated in [95, Part XI].

**Impact.** The consequence is that we have strong evidence that the musical shaping of time is importing the analytical semantics of a compositions into the performance of time. In other words, *musical time turns out to be a construction of a performative time as an expression of musical insights and not simply the projection of a linear symbolic time line.* This also means that the analytical events, harmonic, melodic, and rhythmical in nature, are the analytical constructors of musical time. Other constructors of gestural or emotional ontology are also present, however, they might be more difficult to be boiled down to a concrete formula.

– $\Theta$ –

# Part V

# Conclusions

# 17

# Musical Time Constructs, the Art of Time and Human Creativity

## 17.1 Time Constructs as a Deeply Musical Endeavor

**Summary.** We summarize the musical relevance of time construction.

$-\Sigma-$

This book demonstrates that human time construction has a strong musical character. This musical character may be observed from many phenomena, from child development to African social rhythm (Section 4.4.4) in everyday's life to the event-driven construction of time in Chinese culture (Section 4.4.2.1) or linguistic theories (Section 4.1). It also includes the neuro-pathologies of Alzheimer's (Section 5.2.1.2) and Parkinson's (Section 5.2.1.1) diseases, where musical time adds deep existential significance to patients' lives.

From the very beginning of cultures, time was understood as a double phenomenon: passively given "chronos" and actively lived "kairos." Whereas the former played an important role in physics, the latter became increasingly dominant for the human reality of time. As a such it was recognized as a "dancing presence", which, in Kant's understanding (Section 2.3) is *a priori* internalized in human consciousness. And developing this initial thesis, it becomes an individually embodied existential quality, if not *the* very core of our existential expressivity in Valéry's (Section 2.4) and Merleau-Ponty's (Section 2.10) philosophies.

The musicality of our social and individual lives turns out to be strongly related to the embodied action of our time construction. Following Heidegger (Section 2.10), one may even second his claim that "the sense/meaning of being is temporality." And in so doing, music would then be a most important motor for the construction of our being's sense/meaning.

## 17.2 Art of Time

**Summary.** As artists of time we may give our existence a Heideggerian substance.


G. Mazzola et al., *Making Musical Time*, Computational Music Science,
https://doi.org/10.1007/978-3-030-85629-8_17

– $\Sigma$ –

The gesturally embodied construction of time is much less an administrative objective business than an artistic endeavor. The making of time in its dancing musicality positions itself in the center of human existence, and not objective factuality.

This consequence of our investigations in this book exhibits temporal unfolding as an art that is not a simple ornament of more substantial other topics, but a fundamental activity aiming at touching the in-depth dimension of our lives.

As artists of time we may give our existence a Heideggerian substance without digressing into abstraction from our embodied lives. This restitution of the body as the gate to existential core activity transforms us into artists without any academic program, but focused on the sensual temporality of the art of being, or, rather: existing, stepping out from simply being into a resonant dialogue with ourselves.

## 17.3 Humans Create their Own Time

**Summary.** It is all about the creation of time.

– $\Sigma$ –

The big picture resulting from our reflections is an existential perspective embodied in human time construction, which is a performative instrumental action. While this may sound very philosophical, we insist that it is much more musical than manifestly philosophical. It is not about the passion or love of wisdom, it is the performing realization of wisdom without that destructive masochistic tension of the philosophical tradition of eternal imperfection. Making musical time is feasible, concretely and explicitly, when *making* music. Music is not that far-out ideal, that detached Platonic idea, whose shadow irritates our existence. Yes, we can make time, we can exist without the ifs and buts, just playing our time.

# References


1. Agawu K: African Rhythm. Cambridge University Press 1995
2. Agawu K: Representing African Music. Routledge New York, London 2003
3. Aggleton J P, R W Kentridge, J M Good: Handedness and Musical Ability: A Study of Professioal Orchestral Players, Composers, and Choir Members. Psychology of Music, 22(2), 148-156, 1994
4. Altenmüller E: Neurology of musical performance. Clinical medicine, 8(4), pp. 410-413, 2008
5. Amiot E: Music Through Fourier Space. Springer Series Computational Music Science, Heidelberg, 2016
6. Agustín-Aquino O A, J Junod, G Mazzola: Computational Counterpoint Worlds. Springer Series Computational Music Science, Heidelberg 2015
7. Agustín-Aquino O A, G Mazzola: Contrapuntal Aspects of the Mystic Chord and Scriabin's Piano Sonata No. 5. Submitted for publication 2018
8. Arnett JJ: Adolescents' Uses of Media for Self-Socialisation. J. Youth Adolesc., 24, 519-33, 1995
9. Agawu V K: The African Imagination in Music. Oxford University Press 2016
10. Annan E S: Sankofa art education: A cultural basis for Ghanian art education. Unpublished master's thesis. Department of Art Education. Virginia Commonwealth University, Richmond, VA. 2005
11. Athanasopoulos P and E Bylund: The Whorfian Time Warp: Representing Duration Through the Language Hourglass. Journal of Experimental Psychology, Apr 27 , 2017
12. Bars I: Extra Dimensions in Space and Time (Multiversal Journeys). Springer, Heidelberg 2009
13. Bellman J: Toward a Lexicon for the Style hongrois. Journal of Musicology 9.2: pp. 214-237, 1991
14. Bergson H: Time and freewill: An essay on the immediate data of consciousness (1913). Dover Publications, New York, 2001
15. https://youtu.be/zSB4s9rtj7g
16. Bowling D, Ch Herbst, T Fitch: Social Origins of Rhythm? Synchrony and Temporal Regularity in Human Vocalization. PLOS ONE, Nov. 2013, vol. 8, issue 11
17. Buteau Ch: Motivic Topologies and Their Signification in Musical Motivic Analysis. Masters Thesis, U Laval/Québec 1998




G. Mazzola et al., *Making Musical Time*, Computational Music Science,
https://doi.org/10.1007/978-3-030-85629-8

18. Cavaillès J: Méthode axiomatique et formalisme, Hermann 1938
19. Chapin H et al.: Dynamic Emotional and Neural Responses to Music Depend on Performance Expression and Listener Experience. PLoS ONE, Vol 5, Issue 12, Dec. 2010
20. Cherlin M: Memory and Rhetorical Trope in Schoenberg's String Trio. Journal of the American Musicological Society, 51/3, pp. 559-602, fall 1998
21. Chen L: Guqin Master Chen Leiji. Concert, presented by the Asia Society. `https://www.youtube.com/watch?v=35ZKN-H7at4` 2018.
22. Chernick A, Director: American Masters: Itzhak. Voyeur Films and American Masters Pictures, `https://www.pbs.org/wnet/americanmasters/itzhak-full-film/10786/` 2018
23. Cheyne P, A Hamilton, M Paddison eds.: The Philosophy of Rhythm. Oxford U Press, 2019
24. Chou Wen-Chung. Asian Concepts and Twentieth-Century Western Composers. The Musical Quarterly 57.2: pp. 211-229, 1971
25. Christman S: Handedness in musicians: Bimanual constraints on performance. Brain and Cognition, 22(2), pp. 266-272, 1993
26. Christman S: Eclectic Lefty-Hand: Conjectures on Jimi Hendrix, Handedness, and Electric Ladyland. Laterality 15 (1/2), pp. 253-269, 2010
27. Clayton M: Time in Indian Music. Oxford U Press 2000
28. Clifton T: Music as Heard: A Study in Applied Phenomenology. Yale U Press, New Have 1983
29. Coleman O: Free Jazz. LP, Atlantic 1960
30. Costal F, Am Ockelford, C Shaughnessy: A Report of a Pilot Study Exploring the use of a Mnemonic Strategy for the Recall and use of Functional Language in People with Dementia. J Alzheimers Neurodegener Dis, 6: 042, 2020
31. Cross I: Music as a communicative medium. In: Prehistory of Language, Vol. 1, eds. Botha R and C Knight, Oxford University Press, pp. 113–144, 2009
32. Darwin, Ch: The Origin of Species. New York, NY: Mentor Book 1859/1958
33. Deutsch D: Handedness and Memory for Tonal Pitch. Neuropsychology of Left-Handedness. New York: Academic Press, pp. 263-271, 1980
34. DeNora T: Music into Action: Performing Gender on the Viennese Concert Stage, 1790-1810. Poetics, 30(1-2), pp. 19-33, 2002
35. Domenici C: His master's voice: a voz do poder e o poder da voz. Revista do Conservatório de Música da UFPel, 5, pp. 65-97, 2012
36. Droit-Volet S: Time perception in children: A neurodevelopmental approach. Neuropsychologia 51 pp. 220–234, 2013
37. Engel J: Uncommon Sound: The Left-handed Guitar Players that Changed Music. Left Field Ventures, 2006
38. Eckhardt M, Mueller R, Walker, A. "Liszt, Franz". Grove Music Online. Oxford University Press, 2001
39. Frassinetti F, M Cappelletti, and D Bueti: The Neurobiology of Time Processing. neural aPlasiticiy, Vol. 2016, Article ID 1706373, 2016
40. Fineberg J: Guide to the Basic Concepts and Techniques of Spectral Music. Contemporary Music Review 19/2, 81-113, 2000
41. Flink R: Repeating Ourselves: American Minimal Music as Cultural Practice. U of California Press 2005

42. Freitas S: Modelagem como estratégia para o desenvolvimento de recursos expressivos na performance pianística: três estudos de casos. Universidade Federal do Rio Grande do Sul, 2013
43. Genette, G: Narrative Discourse. An Essay on Method. Cornell UP, Ithaca 1972/1980
44. Goedel K: https://www.thegreatcoursesdaily.com/kurt-godel-wanted-to-revise-our-concept-of-time
45. Good J M M, J P Aggleton, R W Kentridge, J G M Barker, N J Neave: Measuring Musical Aptitude in Children: On the Role of Age, Handedness, Scholastic Achievement, and Socioeconomic Status. Psychology of Music, 25(1), pp. 57-69, 1997
46. Gottlieb G: 'Back-to-front' Bassists: Idiodextrous Approaches to the Double Bass in Jazz (Doctoral dissertation, Middlesex University) 2016
47. Habibi A et al.: Neural correlates of accelerated auditory processing in children engaged in music training. Developmental Cognitive Neuroscience, 21, pp. 1-14, 2016
48. El Haj M and D Kapogiannis: Time distortions in Alzheimer's disease: a systematic review and theoretical integration. npj Aging and Mechanisms of Disease (2016) 2, 16016
49. Hara K: Joe Hisaishi's Soundtrack for My Neighbor Totoro. Bloomsbury Academic, 2020
50. Harrington D L, J L Zimbelman, S C Hinton, S M Rao: Neural modulation of temporal encoding, maintenance, and decision processes. Cerebral Cortex, 20, pp. 1274-1285, 2010
51. Granet M: la Pensée Chinoise. Albin Michel, Paris 1936
52. Grant R M: Beating Time & Measuring Music in the Early Modern Era. Oxford Studies in Music Theory 2014
53. Hall E T: The Dance of Life: The Other Dimension of Time. Doubleday/Anchor Books, New York 1983
54. Hawking S: A Brief History of Time. Bantham Books, New York 1988
55. Hecquet S and A Prokhoris: Fabriques de la danse. PUF, Paris 2007
56. Heifetz J: Brahms' Hungarian Dance #17 in F-Sharp Minor, arranged by Fritz Kreisler. Heifetz (Various Works). SME. 2011, https://www.youtube.com/watch?v=MG5H5lAEvyI
57. Hjelmslev L: Prolégomènes á une théorie du langage. Minuit, Paris 1966
58. Husserl E: Vorlesungen zur Phänomenologie des inneren Zeitbewusstseins. De Gruyter; Auflage: 3rd Edition. Unrevised reprint of the 1st edition 1928 (2000)
59. https://www.cs.hmc.edu/ keller/jazz/improvisor
60. http://www.ixi-audio.net/content/software.html
61. Jackendoff R and Lerdahl F: A Generative Theory of Tonal Music. MIT Press, Cambridge MA, 1983
62. https://www.youtube.com/watch?v=rJVhIbLX0rs
63. Jullien F: La pensée chinoise: En vis-à-vis de la philosophie. Folio, Essais, 2019
64. Kalkandjiev Z: The Influence of Room Acoustics on Solo Music Performances: An Empirical Investigation. Logos Verlag, Berlin, 2015
65. Kant I: Kritik der reinen Vernunft. Meiner, Hamburg 1956
66. Kinderman W: Artaria 195. University of Illinois Press, Urbana and Chicago 2003

67. Klee P: Malerei und Musik. Prestel (2001)
68. Komparu K: The Noh Theater. Waterhill, New York 1983
69. Kopiez R, N Galley, J I Lee: The Advantage of a Decreasing Right-Hand Superiority: The Influence of Laterality on a Selected Musical Skill (Sight Reading Achievement). Neuropsychologia, 44(7), pp. 1079-1087, 2006
70. Kopiez R, N Galley N, A C Lehmann, A C: The Relation between Lateralisation, Early Start of Training, and Amount of Laterality. Neuropsychologia 15(4), pp. 385-414, 2010
71. Kopiez R, H C Jabusch, N Galley, J C Homann, A C Lehmann, E Altenmüller: No Disadvantage for Left-Handed Musicians: The Relationship between Handedness, Perceived Constraints and Performance-Related Skills in String Players and Pianists. Psychology of Music, 40(3), pp. 357-384, 2011
72. Kramer J D: The Time of Music. Schirmer, New York 1988
73. Krebs H: Fantasy Pieces: Metrical Dissonance in the Music of Robert Schumann. Oxford u press 1999
74. Kreisler F: Brahms' Hungarian Dance #17 in F-Sharp Minor, arranged by Fritz Kreisler. The Kreisler Collection—The Berlin HMV Recordings 1926-27. Biddulph, 2020, https://www.youtube.com/watch?v=FCeoOgFVYAA
75. Lewin D: Generalized Musical Intervals and Transformations (1987). Cambridge University Press 1987
76. Kendon A: Gesture: Visible Action as Utterance. Cambridge University Press, Cambridge 2004
77. Kerer M et al.: Explicit (Semantic) Memory for Music in Patients with Mild Cognitive Impairment and Early-Stage Alzheimer's Disease. Experimental Aging Research: An International Journal Devoted to the Scientific Study of the Aging Process, Nov. 2013
78. Kinderman W: Artaria 195. University of Illinois Press, Urbana and Chicago 2003
79. Lerner P: Left Handed Violin? Violinist.Com. 2006. https://www.violinist.com/discussion/archive/8328
80. London J: Hearing in Time. Oxford U Press 2012
81. Lubet A: Indeterminate Origins: A Cultural Theory of American Experimental Music. In: Perspectives on American Music since 1950, edited by James R Heintze, Garland, New York 1999
82. Lubet A: Tunes of impairment: An Ethnomusicology of Disability. Review of Disability Studies: An International Journal 1 (1): pp. 133– 146, 2004
83. Lubet, A. Music, Disability, and Society. Temple University Press, 2011
84. https://www.youtube.com/watch?v=LvgK-abCtQk
85. https://www.youtube.com/watch?v=6nhEgkoByDo
86. https://www.youtube.com/watch?v=sP-IL5vOFps
87. https://imagine.gsfc.nasa.gov/features/yba/M31_velocity/spectrum/doppler_more.html
88. https://www.youtube.com/watch?v=r-GjvtF9EMI&list=OLAK5uy_mVBwnLPHDUbFu6iAVPuuuAxBrbuyGsRkO&index=11&t=0s
89. https://www.youtube.com/watch?v=zCkXWnNaUns&list=OLAK5uy_mVBwnLPHDUbFu6iAVPuuuAxBrbuyGsRkO
90. Macar F, H Lejeune, M Bonnet, A Ferrara, V Pouthas, F Vidal, P Maquet: Activation of the supplementary motor area and of attentional networks during temporal processing. Experimental Brain Research, 142 (4), pp. 475-485, 2002

91. Mazzola G and E Hunziker: Ansichten eines Hirns. Birkhäuser, Basel 1990
92. Mazzola G and O Zahorka: Tempo Curves Revisited: Hierarchies of Performance Fields. Computer Music Journal **18**, No. 1, 1994
93. Mazzola G et al.: The Topos of Music—Geometric Logic of Concepts, Theory, and Performance. Birkhäuser, Basel et al. 2002
94. Mazzola G et al.: The Topos of Music I: Theory. (First volume of second edition of [93]) Springer, Heidelberg 2018
95. Mazzola G et al.: The Topos of Music II: Performance. (Second volume of second edition of [93]) Springer, Heidelberg 2018
96. Mazzola G et al.: The Topos of Music III: Gestures. (Third volume of second edition of [93]) Springer, Heidelberg 2018
97. Mazzola G et al.: The Topos of Music IV: Roots. (Fourth volume of second edition of [93]) Springer, Heidelberg 2018
98. Müller, Stefan and Mazzola, Guerino: Constraint-based Shaping of Gestural Performance. Proceedings of the ICMC 03, ICMA, Ann Arbor 2003
99. Mazzola G and P B Cherlin: Flow, Gesture, and Spaces in Free Jazz—Towards a Theory of Collaboration. Springer Series Computational Music Science, Heidelberg 2009
100. Mazzola G: Musical Performance—A Comprehensive Approach: Theory, Analytical Tools, and Case Studies. Springer Series Computational Music Science, Heidelberg, December 2010
101. Mazzola G, J Park, F Thalmann: Musical Creativity. Springer, Heidelberg 2011
102. Mazzola G, M Mannone, Y Pang: Cool Math for Hot Music. Springer, Heidelberg 2016
103. Mazzola G, M Mannone, M O'Brien, Y Pang, N Torunsky: All About Music. Springer, Heidelberg 2016
104. Mazzola G et al.: The Future of Music. Springer, Heidelberg, 2020
105. Mazzola G et al.: Basic Music Technology. Springer Series Computational Music Science, Heidelberg, 2018
106. Mazzola G: Functorial Semiotics for Creativity. Journal of Mathematics and Music 2020, Vol. 14, no. 1, 2019
107. Merleau-Ponty M: Phénoménologie de la perception. Gallimard 1945
108. Merleau-Ponty M: Phenomenology of Perception. Routledge 1958
109. McTaggart M E: The Unreality of Time. Mind 17: 457–73, 1908
110. Menuhin Y: Brahms' Hungarian Dance #17 in F-Sharp Minor, arranged by Joseph Joachim. 100 Best Menuhin. Warner Classics, 2009
111. Michaels, K et al.: 'Dance Therapy' as a Psychotherapeutic Movement Intervention in Parkinson's Disease. Complementary Therapies in Medicine, pp. 248–252, Oct. 2018
112. Miller G: Evolution of Human Music through Sexual Selection. In: The Origins of Music (Ed. Wallin NL, Merker B, Brown S), 329-60, The MIT Press, Cambridge 2015
113. Miracle A, J D Yapita:. Time and space in Aymara. In M. J. Hardman (Ed.), The Aymara language in its social and cultural context (pp. 33–56). Gainesville: University Presses of Florida, 1981
114. Morishita Ch: A Musical Engagement with 'MA'—Japanese Aesthetics of Space and Time. Diss. University of Huddersfield, 2011
115. Moten F: In the Break: The Aesthetics of the Black Radical Tradition. University of Minnesota Press, Minneapolis 2003, p. 39

116. Motoyasu H et al.: Dysfunctional counting of mental time in Parkinson's disease. Scientific Reports, Volume 6, id. 25421, 2016
117. Munyaradzi G and W Zimidzi: Comparison of Western Music and African Music. Creative Education,, vol. 3, No. 2, 193-195, 2012
118. Muller T and A C Nobre: Perceiving the passage of time: neural possibilities. Ann. N.Y. Acad. Sci. ISSN 0077-8923, 2014
119. Newton I: Philosophiae Naturalis Principia Mathematica (1726). University of California Press 2016
120. Núñez R E, E Sweetser: With the Future Behind Them: Convergent Evidence From Aymara Language and Gesture in the Crosslinguistic Comparison of Spatial Construals of Time. Cognitive Science, 30, pp. 401-450, 2006
121. Oldfield R C: The Assessment and Analysis of Handedness: The Edinburgh Inventory. Neuropsychologia 9.1: 97-113, 1971
122. `https://youtu.be/TK1Ij_-mank`
123. Pang Y: Rural and Urban Explorations of Transnational Performance: A Dialogue Between Traditional Southwestern Chinese and Contemporary North American Music & Dance. 2018 Retrieved from the University of Minnesota Digital Conservancy, `http://hdl.handle.net/11299/202159`
124. Pedelty M, R Dirksen, T Hatfield, Y Pang, E Roy: Field to Media: Applied Ecomusicology in the Anthropocene. Popular Music 39, no. 1 (2020): pp. 22–42. doi:10.1017/S0261143019000540
125. Pereira A S et al.: Chimpanzee lip-smacks confirm primate continuity for speech-rhythm evolution. The Royal Society Journal, Biology Letters, 27 May 2020
126. Peters M: Handedness: Effect of Prolonged Practice on Between Hand Performance Differences. Neuropsychologia, 19(4), pp. 587-590, 1981
127. Piaget P: Le développement de la notion de temps chez l'enfant. Presses Universitaires de France, Paris 1946
128. Piro J and C Ortiz: No Association between Music Ability and Hand Preference in Children. Journal of Motor Behavior, 42(5), pp. 269-275, 2010
129. Pyykönen K: A Handful of Considerations Perspectives on Left-handedness in Violin Playing and Violin Pedagogy. Thesis, Masterprogram i Musikpedagogik, Institution för Musik, Pedagogik och, Samhälle, 2015
130. Puttka M: Learning to Dance Means Learning to Think! In: Puttke M et al. (Eds.): The Neurocognition of Dance. Psychology Press, Hove and New York 2010
131. Quinan C: A Study of Sinistrality and Muscle Coordination in Musicians, Iron Workers and Others. Archives of Neurology & Psychiatry, 7(3), pp. 352-360, 1922
132. Ravignani A, B Gingras, R Asano, R Sonnweber, V Matella'n, W T Fitch: The evolution of rhythmic cognition: New perspectives and technologies in comparative research. Proceedings of the Annual Meeting of the Cognitive Science Society, 35, pp. 1199–1204, 2013
133. Ravignani A, D Bowling, S Kirby: The psychology of biological clocks: a new framework for the evolution of rhythm. Paper presented at: The Evolution of Language: Proceedings of the 10th International Conference; Vienna, Austria, April 14-17, 2014
134. Ravignani A: Timing of Antisynchronous Calling: A Case Study in a Harbor Seal Pup (*Phoca vitulina*). Journal of Comparative Psychology 2018
135. Ravignani A and and K de Reus: Modelling Animal Interactive Rhythms in Communication. Evolutionary Bioinformatics, vol. 15, pp. 1-14, 2019

136. Ravignsni A et al.: Rhythm in speech and animal vocalizations: a cross-species perspective. Ann. N.Y. Acad. Sci. XXX, pp. 1-12, 2019
137. Ravignani A, M Larsson, J Richter: Bipedal Steps in the Development of Rhythmic Behavior in Humans. Music and Science, vol. 2: pp. 1-14, 2019
138. Repp B: Sensimotor Synchronization: A Review of the Tappiing Literature. Psychonomic Bulletin & Review 12 (6): pp. 969-92, 2005
139. Riemann H: System der musikalischen Rhythmik und Metrik. Breitkopf und Härtel, Leipzig 1903
140. Schenker H: The Art of Performance. Heibert Esser (Ed.), Irene Schreier Scott (Translation), Oxford U Press 2000
141. Schulkin J and G B Raglan: The evolution of music and human social capability. Frontiers in Neuroscience, Sept. 2014
142. Sessions R: Questions About Music. W. W. Norton and Company, 1971
143. `http://www.chinatourguide.com/chengdu/sichuan_opera.html`
144. Shepp A: Coral Rock. America Records 1970
145. Sideways: Why Miyazaki's Films Sound Pretty. `https://youtu.be/4bZ19hnr8vc`
146. Sinha Ch and P Gädenfors: Time, space, and events in language and cognition: A comparative view. Annals of the New York Academy of Sciences, August 2014
147. Smits R: The puzzle of left-handedness. Reaktion Books, 2011
148. Smolin L: Time Reborn. Mariner Books, Boston, New York 2020
149. Snowman J: The Left and Right Hands of the Eighteenth-Century British Musical Prodigies, William Crotch and Samuel Wesley. Laterality, 15(1-2), pp. 209-252, 2010
150. Soares S, B V Atallah, J Pato: Midbrain Dopamine Neurons Control Judgement of Time. Science 09 Dec. Vol. 354, Issue 6317, pp. 1273-1277, DOI: 10.1126/science.aah5234, 2016
151. Stegmoller E L et al.: Experiences of Persons With Parkinson's DiseaseEngaged in Group Therapeutic Singing. Iowa State University 2017
152. `https://www.youtube.com/watch?v=lOGN4OEL1VU`
153. Su Y-H, Salazar-López E: Visual Timing of Structured Dance Movements Resembles Auditory Rhythm Perception.
154. Taruskin R: Text and Act: Essays on Music and Performace. New York: Oxford University Press, 1995
155. Taylor C: Burning Poles. VHS, Mystic Fire 1991
156. Thompson J (ed.): Taiyin Daquanji (Great Collection of Superlative Sound). `http://www.silkqin.com/02qnpu/05tydq/ty5.htm#p12` N.d.
157. Tokita A McQ, D Hughes: Context and change in Japanese music. Ashgate Publishing Limited, 2007
158. Uhde J and R Wieland: Forschendes Üben. Bärenreiter 2002
159. Ustun B:"Time is Production": Process-Art, and Aesthetic Time in Paul Valéry's Cahiers. Humanities 7, 4; doi:10.3390/h7010004, 2018
160. Valéry P: Philosophie de la danse (1938). Oeuvres I, Variété, "Théorie poétique et esthétique", 1390-1403, Gallimard, Paris 1957
161. Valéry P: Cahiers I-IV (1894-1914). Celeyrette-Pietri N and Robinson-Valéry J (eds.), Gallimard, Paris 1987
162. Wahl J: Traité de métaphysique, cours sorbonne. Payot, Paris 1968
163. Wang, G: Interlopers in the Realm of High Culture: "Music Moms" and the Performance of Asian and Asian American Identities. American Quarterly 61.4: pp. 881-903, 2009

164. Wendorff R: Zeit und Kultur. Westdeutscher Verlag, Opladen 1980
165. Whorf B L: Language, Thought and Reality. The Theosophist, Madras 1941
166. Wiener M, et al.: The image of time: A voxel-wise meta-analysis 60–71. j.neuroimage.2009.09.064
167. Pang Y: (1989) Philia in Melody. Youtube. Uploaded by Pang, 9 September 2012, https://www.youtube.com/watch?v=P1ivluW2rSE.
168. Yang M: East Meets West in The Concert Hall: Asians and Classical Music in the Century of Imperialism, Post-Colonialism, and Multiculturalism. Asian Music 38.1: pp 1-30, 2007
169. Yen Pang's website: https://www.yanpangcreate.com
170. Yourgru P: A World Without Time. Basic Books 2005
171. Yust J: Organized Time. Oxford U Press 2018
172. Zoon J: Density 21.5. Edgard Varèse: The Complete Works. CD. Decca London Polygram 289 460 208–2, https://www.youtube.com/watch?v=s6-TBeDL1nA 1998

# Index




G. Mazzola et al., *Making Musical Time*, Computational Music Science,
https://doi.org/10.1007/978-3-030-85629-8